The Privilege of
Blue Skies

Tracing the Uneven Legacy of
Air Pollution Across Nations

WORLD SCIENTIFIC SERIES IN MANAGEMENT STRATEGIES, POLICIES AND IMPLEMENTATION

Series Editor: Gideon Markman *(Colorado State University, USA, Gent University, Belgium & Audencia Business School, France)*

Published

Vol. 1   *The Privilege of Blue Skies: Tracing the Uneven Legacy of Air Pollution Across Nations*
by Saheli Nath

World Scientific Series in Management Strategies, Policies and Implementation

The Privilege of Blue Skies

Tracing the Uneven Legacy of Air Pollution Across Nations

Saheli Nath

University of Central Oklahoma, USA

World Scientific

NEW JERSEY • LONDON • SINGAPORE • BEIJING • SHANGHAI • TAIPEI • CHENNAI

Published by

World Scientific Publishing Co. Pte. Ltd.
5 Toh Tuck Link, Singapore 596224
USA office: 27 Warren Street, Suite 401-402, Hackensack, NJ 07601
UK office: 57 Shelton Street, Covent Garden, London WC2H 9HE

Library of Congress Control Number: 2025013240

British Library Cataloguing-in-Publication Data
A catalogue record for this book is available from the British Library.

World Scientific Series in Management Strategies, Policies and Implementation — Vol. 1
THE PRIVILEGE OF BLUE SKIES
Tracing the Uneven Legacy of Air Pollution Across Nations

ISBN 9789819810932 (hardcover)
ISBN 9789819810949 (ebook for institutions)
ISBN 9789819810956 (ebook for individuals)

For any available supplementary material, please visit
https://www.worldscientific.com/worldscibooks/10.1142/14248#t=suppl

Desk Editors: Aanand Jayaraman/Venkatesh Sandhya

Typeset by Stallion Press
Email: enquiries@stallionpress.com

Dedicated to

My parents, who brought me to this beautiful world,
And to all of you who strive hard to keep it green —
You write Earth's future in bedtime tales you tell your children
And in your actions that will long outlast these mundane pages.

Foreword

The air we breathe — so essential, so constant, and so taken for granted — is becoming one of the most contested resources of our time. Dr. Saheli Nath's compelling book offers a timely, comprehensive, and much-needed analysis of the complex forces shaping air pollution in the 21st century. Through a blend of historical perspective, analytical rigor, and human narrative, this book unveils a sobering truth: While many of us enjoy the "blue skies," this privilege often comes at the expense of communities and ecosystems in other parts of the world. By tracing the economic, technological, and sociopolitical pathways through which pollution is displaced from one region to another, Dr. Nath presents a deeply interconnected view of the global air pollution crisis.

At the heart of this book is a powerful insight: Air pollution is not merely a byproduct of industrialization — it is also a reflection of the global systems of production, consumption, and environmental governance. Dr. Nath deftly demonstrates how, following environmental disasters like the Donora Smog of 1948 and London's Great Smog of 1952, stricter air quality regulations in developed nations triggered the migration of pollution-intensive industries to developing countries. While these regulatory shifts afforded blue skies to London, Pittsburgh, and Los Angeles, they simultaneously darkened the skies of industrial hubs in India, China, and Southeast Asia. This shift, described in the book as the creation of a "privilege of blue skies," exposes an uncomfortable reality: The cleaner air enjoyed by one part of the world often comes at the expense of another.

The book does not stop at historical analysis. It also addresses the modern solutions and technological innovations that are widely touted as pathways to cleaner air. In her sharp critique, Dr. Nath explores the unintended consequences of "green" and "smart" technologies — like cloud computing, smart cities, and energy-efficient infrastructure — revealing how they often conceal hidden costs. For example, while cloud computing is marketed as a "clean" alternative to physical infrastructure, the vast network of data centers powering the cloud requires enormous amounts of electricity and cooling, adding to pollution in regions where data centers are clustered. Similarly, the drive to build "green" cities may appear sustainable on the surface, but the displacement of vulnerable communities and the environmental costs of construction often remain hidden from public scrutiny. By exposing these paradoxes, Dr. Nath encourages readers to look beyond surface-level solutions and scrutinize the deeper, systemic impacts of environmental interventions.

One of the most profound contributions of this book is its exploration of emerging solutions from across the world. While much of the existing literature on air pollution focuses on technological advances or regulatory frameworks, Dr. Nath introduces a more multidimensional approach. She highlights the potential of Indigenous knowledge systems and grassroots movements, demonstrating that solutions to air pollution are not confined to boardrooms or scientific labs. For instance, the fire management practices of Indigenous Australian communities, the grassroots environmental movements in India, and the local air quality initiatives in Mexico City all reveal that practical, human-centered solutions can arise from unexpected sources. These examples point to a larger insight: Successful responses to air pollution require both top-down policy interventions and bottom-up social action.

The book also highlights the role of corporate–civic partnerships, offering a vision of how businesses, governments, and local communities can co-create solutions. Unlike purely market-driven approaches, these partnerships balance the profit motives of firms with the public good, resulting in more equitable outcomes. By examining sustainability initiatives and environmental, social, and governance (ESG) criteria, this book shows how firms are increasingly held accountable for their environmental impact. However, Dr. Nath does not offer blind praise for these initiatives; she carefully dissects their limitations, ensuring readers remain aware of the potential for "greenwashing" — the misrepresentation of corporate practices as sustainable.

Perhaps the most evocative contribution of this book is its human dimension. Interwoven throughout the analysis are vivid personal stories and visceral images that remind us of what is truly at stake. The narrative is bookended by Dr. Nath's recollection of her childhood in Nalhati, India, a village where she experienced the fresh breezes of the evening and gazed at the clear night sky. In one of the most poignant sections of the book, she contrasts these childhood memories with the modern-day realities of Delhi, where the sky is now a persistent gray, the air carries the acrid smell of exhaust fumes, and rivers like the Yamuna flow with toxic white foam. These images are not mere anecdotes; they embody the heart of the book's argument — air pollution is not an abstraction. It shapes our lives in concrete and deeply human ways.

The final chapters of the book present an urgent call for action. But unlike simplistic calls for "sustainability" or "green living," this call is grounded in the logic of collective action and systemic change. Dr. Nath reminds us that while the problem of air pollution may seem vast and overwhelming, human resilience, creativity, and persistence can still inspire change. Just as the fire ants she watched as a child displayed discipline and unity, humanity too can confront the challenge of air pollution with similar diligence and persistence.

This book is an invitation to think differently about air. It moves beyond technical debates about emissions levels and pollutants to reveal the deeper, often hidden dynamics of power, privilege, and displacement. It calls for a rethinking of what it means to "make it" in modern society — not just in terms of material wealth or urban progress, but in terms of the air we breathe, the health we sustain, and the dignity we preserve. This shift in perspective, as Dr. Nath argues, is essential for a world where clean air is recognized as a fundamental human right.

Few books manage to combine academic rigor with human storytelling so seamlessly. Dr. Nath has crafted a work that is as intellectually rich as it is emotionally resonant. It will challenge policymakers, scholars, activists, and citizens to reflect on the deeper costs of progress and to reconsider what it means to "make it" in life. It will inspire readers to ask new questions: Who pays the price for progress? Who benefits? And how can we chart a better course forward?

This book is not merely a chronicle of the past, nor is it a distant warning about the future. It is a profound reminder that the struggle for clean air is happening right now — in cities like Delhi, Beijing, Los Angeles, and Jakarta. Through clear-eyed analysis, vivid narrative, and forward-looking solutions, Dr. Saheli Nath has given us not just a book, but a guidepost for navigating the complex airscapes of the 21st century.

Breathe deeply. Read slowly. And be ready to think differently about the air around you.

Gideon D. Markman

Preface

I initially took pen to paper to write this book to explore the history of environmental sustainability and the changing role of businesses in this conversation over time. But as I really started to write and research on the topic, I found the fundamental issue at hand was human nature. Our natural drive for progress, comfort, and prosperity has led us down a somewhat unexpected path of ecological unraveling. For example, in many developing countries like China and India, coal mining was once viewed as a major vehicle of modernization. For numerous villages without electricity in the 1980s, coal-powered electricity cleared the path to a higher quality of life (Chou, 1979). Coal miners were our everyday heroes who pushed the country toward development through hard, risky labor — the flame of their lives burning to transform society. Today, coal mining is frowned upon as a major air polluter, and governments around the world are eager to explore viable cleaner alternatives. Once hailed, now reviled — how the wheels of progress have turned on coal. The miners themselves, once celebrated as stalwart champions of progress who could hold their heads high, now find their legacy dimmed by time's shifting shadows, their sacrifices recast not as noble steps toward modernization but as unwitting contributions to our planet's decline.

As I delved deeper into the subject, I was struck by the complexity of emotions it evoked. On the one hand, I could feel love and hope, infused with a sense of positivity and peace, when I read how hard entrepreneurs, educators, environmentalists, and scientists all across the globe have worked to understand, analyze, and address the issue of air pollution. I could feel the resonance of life when I witnessed the incredible resilience and ingenuity of communities rising to meet environmental challenges. From grassroots movements in developing countries to technological innovations by different companies, we as human beings have shown a remarkable capacity for adaptation and problem-solving.

Yet, I could not ignore the darker aspects of our nature — our inclination to prioritize short-term gains over long-term sustainability, our ability to distance ourselves from the consequences of our actions. For instance, in 2022, the world produced approximately 62 million tonnes of electronic waste (WHO, 2024), most of which were disposed of in countries like China, India, Ghana, Ivory Coast, Benin, and Liberia. Even as we rush to the malls to purchase the latest smartphone, thousands of workers in Accra, Ghana, sort through mounds of burning electronic waste in hazardous conditions, often afflicted with burns, infected wounds, back pain, respiratory illnesses, and debilitating headaches (Yeung, 2019). Thinking of all the pain suffered, the tears shed, and the pressure endured in the past years by people in these vulnerable communities, I could feel a nameless rage in my heart when I considered the scale of the crisis we face and the slowness of our response.

Throughout this journey of research and writing, I've come to realize the profundity of Gong Ji-Young's (2009/ 2023) *Togani (The Crucible)* — that many of us are not fighting to change the world, but rather fighting to keep the world from changing us. It is a constant struggle to maintain hope and drive positive change in the face of overwhelming challenges. This book is a testament to that struggle — an exploration of how human nature both creates and solves environmental problems. In these pages, you'll find a narrative that spans continents and decades, from the smog-filled cities of the Industrial Revolution to the tech-driven solutions of the 21st century. We will examine the interplay between economic development and environmental protection, the role of policy and technology in addressing air pollution, and the power of individual and collective action.

My hope is that this book will not only inform but also inspire. By understanding the roots of our environmental challenges and the innovative ways we are addressing them, perhaps we can chart a course toward a more sustainable future — one that aligns our economic aspirations with the health of our planet and the air we breathe. As you read, I invite you to reflect on your own role in this global narrative. For in the end, the story of air pollution — and its solution — is the story of all of us, our choices, our actions, and our shared future on this planet we call home.

References

Chou, S. H. (1979). Industrial modernization in China. *Current History*, *77*(449), 62–85.

Gong, J. Y. (2009/2023). *Togani (B. Fulton & J. Fulton, Trans.)*. Honolulu, HI: University of Hawaii Press.

WHO. (2024, October 1). Electronic waste (e-waste). *World Health Organization*. Retrieved on October 5, 2024 from https://www.who.int/news-room/fact-sheets/detail/electronic-waste-(e-waste).

Yeung, P. (2019, May 29). The toxic effects of electronic waste in Accra, Ghana. *Bloomberg*. Retrieved on February 21, 2024 from https://www.bloomberg.com/news/articles/2019-05-29/the-rich-world-s-electronic-waste-dumped-in-ghana.

About the Author

Saheli Nath is a vocal advocate for using business as a force for social and environmental good. In 2024, she was awarded the "Pushing the Boundary Award" by the *Strategizing Activities & Practices Division* of the *Academy of Management*, a premier international professional association for scholars in the discipline of management and organizations. She obtained her doctorate from the Kellogg School of Management at Northwestern University, USA, and her M.Phil. from the University of Cambridge in England. As an interesting side note, Dr. Nath participated in competitive chess when she was young, representing India in various World, Asian, and Commonwealth Youth Chess Championships. She was also the only female member of the 2013 winning University of Cambridge team in the *131st Oxford vs. Cambridge Varsity Chess Match*, which represents "the oldest fixture in chess, played annually since 1873 apart from war years" (Barden, 2024). Along with her academic research, Dr. Nath's extensive traveling experiences from her childhood and interactions with people from various countries and cultures shaped many of the views presented in this book.

Reference

Barden, L. (2024, March 4). Chess: Oxford vs Cambridge almost level after 151 years. *Financial Times*. Retrieved on October 5, 2024 from https://www.ft.com.

Acknowledgments

I am truly grateful to Dr. Monica Prasad, whose visionary problem-solving sociology workshops have been a source of inspiration since my days as a doctoral student at Northwestern University. This endeavor has brought together like-minded colleagues from around the world who work closely to address complex challenges. I was delighted to have the opportunity to present a summary of the early draft of my book at the "Solving Global Poverty" workshop in the summer of 2024, where I received invaluable feedback on improving the content of my work. Through rigorous discussions and constructive critiques from fellow participants, I was able to refine my arguments, strengthen my analytical framework, and identify crucial gaps in my initial research.

Furthermore, I deeply appreciate the help of Dr. Ana Maria Peredo, who gave me the opportunity to be closely involved with the *International Academy of Research in Indigenous Management and Organizational Studies*. During my conversations with representatives of Indigenous communities and their allies from around the world, I gained a better understanding of how Indigenous knowledge of the land, passed down through generations, offers invaluable insights that can help mitigate environmental impacts and promote sustainable development. I learned more about the valiant struggle of Indigenous communities across the globe to ensure that their rights are respected and their consent is obtained before any corporate or government development projects take place on their ancestral lands.

I would also like to express my sincere gratitude to Jessica Thomas, co-founder of *B Academics*, and Dr. Kristin Joys, the Director of the Social Impact & Sustainability Initiative at the University of Florida.

They kindly facilitated my involvement with *B Academics* as a member of its Board of Directors, a role which provided me with numerous insights. The organization's focus on fostering changemakers who view business as a force for social good instead of a tool to maximize profits and our discussions on strategies, practices, and resources to support this mission were particularly enlightening. These conversations instilled in me a renewed sense of hope and purpose, especially when confronting the complex issue of air pollution addressed in this book.

In addition, I would like to extend my heartfelt thanks to the Sustainability Center at the University of Central Oklahoma for providing an environment that nurtured my understanding of environmental stewardship and sustainable practices. In particular, I am deeply appreciative of the opportunity to work alongside Dr. Ed Cunliff and Mark Walvoord in organizing the Earth Day Photo Contest and the Earth Week events at our university. The process of curating photographs that capture the beauty of our planet and coordinating events that celebrate our environment allowed me to witness firsthand the power of community engagement in fostering environmental awareness. The participants' passion for sustainability and their innovative approaches to raising ecological consciousness were both inspiring and educational. Through our shared efforts, I gained a deeper appreciation for how seemingly small, everyday actions can collectively contribute to creating a more beautiful and sustainable planet. They have reinforced my belief in the importance of grassroots efforts and individual actions in addressing global challenges.

Finally, I would like to thank my parents for their serendipitous decision to teach me the game of chess when I was a child. This set in motion a series of events that would shape not only my understanding of strategy and competition but also my perception of the world around me. Through their unwavering support, I had the privilege of traveling to chess tournaments across the nation and around the globe. These journeys, ostensibly about moving pieces on a board, became profound lessons in global citizenship and environmental stewardship. As I moved from city to city, country to country, I witnessed firsthand the stark realities of our changing environment. The varying quality of air I breathed — from the crisp, clean atmosphere of some locations to the heavy, polluted haze of others — left an indelible mark on my consciousness. This book is a culmination of that journey — from a child learning to castle on a chessboard to an advocate striving to clear our skies. It stands as a testament to the unexpected ways in which our paths toward awareness and action can begin.

Contents

Foreword   vii

Preface   xi

About the Author   xv

Acknowledgments   xvii

Introduction   xxi

Mass Production, Mass Consumption, and the Vanishing Blue   **1**

Chapter 1   From Factories to Firmament: Offshoring
             Industries Across Borders   3

Chapter 2   Clouded Rainbows: Pricing Consumerist
             Aspirations and Quality-of-Life Pursuits   13

Chapter 3   Pricing the Last Breath: The Global
             Hierarchy of Human Worth   23

Chapter 4   Shades of Inequality: Air Pollution's Uneven
             Toll across Communities   33

Chapter 5   Digital Haze Ahead: The Hidden Environmental
             Cost of Cloud Computing   43

A Tale of Two Countries: Blue Skies Lost and Recovered   **53**

Chapter 6   China's Blue Sky Recovery: A Journey from
              Smog to Sustainability   55

Chapter 7   India's Vanishing Blue Skies: A Tale of
              Growth and Neglect   65

Chapter 8   Clearer Horizons Forward: Developing
              Nations' Solutions and Challenges   73

Chapter 9   Tales of Caution: Green Gentrification and
              the Hidden Cost of Clean Air   83

Chapter 10   Space Race for Clean Air: How Satellite
              Cities Reshape Atmospheric Politics   91

Stewardship: Charting a Path Toward a Global Ethical Skyline **101**

Chapter 11   For Nurrundere: Integrating Indigenous
              Wisdom into Sustainability   103

Chapter 12   Ground-Up Blue: Local Movements
              Against Gray Skies   115

Chapter 13   Corporate–Civic Partnerships: Building
              Bridges for Azure Skies   123

Chapter 14   Flight to B Economy: Redefining Corporate
              Success Through ESG Metrics   133

Chapter 15   The Right to Breathe: Air Quality as the
              Next Human Rights Frontier   141

Epilogue: Shaping Blue Skies for All   149

Glossary   155

Index   163

Introduction

When I try hard to recall my childhood memories, I regrettably find them quite hazy, with days and months blurring into each other like colors in a mottled painting. Yet, there are some sceneries that stand out vividly, akin to a splash of bright ink on jaded black-and-white film. I remember my beloved blue tricycle — I would ride it up a small hill in front of our home and then sit on the sturdy rocks to stare at fire ants. I cannot remember exactly what about the ants so fascinated me — perhaps it was their diligence, discipline, and unity, but maybe it was simply their gorgeous red color. Then, there was the tubewell behind our house — a monkey would sit on it every so often and we would look intently at each other, making fruitless attempts at telepathy. Another lucid picture is that of my mom lighting up the hurricane kerosene lamp. Our home would often lose electricity, and she would light the lamp and carry me outside in her arms. We would cool down together in the light breeze, looking at the bright stars in the clear night sky. Despite its simple pleasures and daily inconveniences, life in the Indian village of Nalhati held a special charm.

I recall especially cherishing the occasional trips to the market with my father. As a child, I had a sweet tooth. My dad would often distract me with the story of *The King and the Tailorbird*, and he would take me down the mud roads to visit the local market to get *rasgullas* made with jaggery. I would marvel at the strong bullocks pulling the carts and exclaim at the occasional *lorries* that passed by. There were hardly any private cars on the roads. Private cars were seen as a privilege of the rich, and to be able to "buy a house and a car" was emblematic of having "made it" in life. In the early 1990s, many Indian parents hoped their child would move from the villages to an urban area, become a doctor or an engineer, contribute to the country, and be able to buy a house and a car.

Today, when I visit Delhi, India's capital, I see the city brightly lit over Christmas. I see hundreds of thousands of private cars. I see construction everywhere. What used to be fields and forests have now given way to large urban apartment complexes, and I feel truly joyous that so many people have "made it." Then, I look up and see the gray haze covering the skies. The vibrant blue-and-white clouds that I saw when traveling to Delhi as a child for my chess competitions are now elusive. I walk outside and feel the temperatures that are hotter than anything I experienced growing up. And then I feel frightened; many of us have "made it," but at what cost?

The cost, I realize, is written in the very air we breathe. The sweet scent of jasmine and plumeria that once permeated the evening air is now overpowered by the acrid smell of exhaust fumes. The horizon, once clearly defined by the silhouette of distant trees and buildings, now blurs into an indistinct gray mass. The Yamuna River, once famous for its serene beauty, now flows sluggishly, its waters covered with an uncanny white foam — a deadly cocktail of untreated sewage, industrial effluents, and chemical detergents. The sunset over the river, when visible, is a muted spectacle — the vibrant oranges of my memory replaced by a dull, reddish glow. Along its banks, in a heartbreaking sight that captures both devotion and desperation, some pilgrims still perform their daily rituals in these poisoned waters, risking serious health issues, their faith unchanged even as their beloved river has transformed beyond recognition. As darkness falls, the city lights up, a dazzling display of human achievement. Yet, I can't help but mourn the loss of the twinkling stars and the gentle glow of the moon — now hidden behind a veil of pollution.

Standing here amid the haze of urban development, breathing in air thick with the cost of economic growth, I'm struck by a profound sense of loss and responsibility. The progress we have made — the cars, the apartments, the glittering malls — suddenly seems hollow in the face of what we have lost. However, amidst this gloom, I see glimmers of hope. I hear of citizens' movements fighting for cleaner air, of young entrepreneurs developing innovative solutions for waste management, and of schools teaching children about environmental conservation. I see rooftop gardens sprouting amidst the concrete jungle and solar panels gleaming on building terraces. Perhaps we can still reclaim our blue skies, our clean rivers, our starry nights. Perhaps we can redefine what it means to "make it" — not in terms of cars and houses, but in terms of the quality of the air we breathe, the water we drink, and the food we eat. I think back to those resilient fire ants of my childhood, their strength lying in their unity and persistence. Perhaps that's the lesson we need to learn. The challenge of cleaning our air may seem insurmountable, but if we approach it with the same diligence and collective effort, there's hope.

When I first began conceiving this book, my initial months of research on air pollution across continents revealed a troubling pattern — the quest for cleaner air in one region often came at the cost of darker skies elsewhere. The story of how this pattern emerged begins in Chapter 1 by examining the historical shift of polluting industries from developed to developing countries, in light of critical events like the Donora Smog of Pennsylvania in 1948 and the Great Smog of London in 1952. It reveals how Western nations, faced with tightening environmental regulations to address adverse events, effectively exported their pollution problems to developing countries eager for economic growth, demonstrating a tendency to address local issues by displacing them rather than solving them at their root. This industrial migration created the "privilege of blue skies" for Western consumers while worsening pollution in developing nations.

This privilege, however, didn't emerge in isolation. It was born from a profound transformation in how we live and consume — a shift that Chapter 2 examines in detail. Chapter 2 explores the shift from a pre-industrial society of simplicity and frugality to our modern "use-and-throw" culture, highlighting how the promise of a better life through industrial progress and economic growth tapped into our deep-seated desire for comfort and security, leading to a societal change that prioritized acquisition and consumption. The chapter examines the unintended consequences of this pursuit of prosperity and consumption-driven lifestyle, particularly the exportation of the environmental burdens associated with "raising our quality of life" to developing nations, revealing our tendency to seek immediate material benefits even at the cost of long-term global well-being.

The true cost of this transformation becomes even more stark when we consider how we use tools in policy analysis to quantify what society is willing to spend on safety improvements that reduce mortality risks. The same breath of air that would trigger legal sanctions in one country might be deemed an acceptable risk in another based on economic models that evaluate collective willingness to pay to save lives differently. Chapter 3 takes a look at how the "Value of Statistical Life" varies dramatically between nations, leading to pollution-related deaths in developing countries being systematically undervalued in cost–benefit analyses. The chapter demonstrates how these differential valuations not only reflect existing global inequalities but actively perpetuate them by making it "economically rational" to export environmental hazards to regions where lives are "statistically" worth less in decision-making models. This economic framework offers a lens to understand the persistent migration of polluting industries to developing nations and challenges us to confront the ethical implications of monetizing human life.

The aforementioned inequity manifests most visibly in how pollution affects different communities, which Chapter 4 explores through the lens of environmental justice. Chapter 4 elucidates the "race-to-the-bottom" phenomenon, where pollution-intensive industries tend to relocate to countries with lax environmental standards, and investigates how our propensity for in-group favoritism and psychological distancing has led to an uneven distribution of environmental burdens, disproportionately affecting poorer communities and vulnerable populations.

Even as we embrace technological solutions to our pollution issues, we find new challenges emerging. Chapter 5 reveals the paradox of our increasingly digital world. While the shift to cloud computing and digital services appears environmentally friendly on the surface, it involves significant adverse impacts. Beginning with an examination of data centers' growing energy consumption, equivalent to many small nations' total usage, this chapter explores how our digital footprint, though seemingly more sustainable, masks its own costs.

The second part of this book analyzes specific cases of serious air pollution in more depth, beginning with two crucial case studies that epitomize the global struggle between economic development and environmental protection. Chapter 6 starts with how China's rapid industrialization, driven by mass production to support global consumption, led to severe air pollution, and subsequently highlights China's multi-faceted approach to combating air pollution, showcasing humanity's capacity for adaptation and problem-solving when faced with urgent environmental challenges. It draws attention to how collective action and political will can lead to significant environmental improvements.

This analysis naturally leads to Chapter 7's examination of India's ongoing air pollution crisis, which provides a striking contrast to China's trajectory. It details the severe health and economic impacts of air pollution in India, with a particular focus on major cities like Delhi.

The comparison between these two Asian giants offers crucial insights into how different governance approaches, technological innovation, and resource allocation strategies can lead to divergent outcomes in the fight against air pollution.

Building on these real-world examples, Chapter 8 shifts focus to practical solutions, synthesizing lessons from both successful and struggling nations to outline a comprehensive approach to addressing air pollution. This chapter bridges the gap between the problem analysis in the earlier chapters and the solution-focused latter half of this book. Chapter 8 emphasizes the need for a multi-faceted approach to address air pollution, which includes energy sector transitions, urban planning reforms, and institutional capacity building. This chapter explores the potential of technologies such as artificial intelligence (AI) and robotics in environmental monitoring, sustainable manufacturing, and policymaking. Finally, it underscores the importance of balancing technological advancements with ethical considerations and equitable implementation to effectively tackle global pollution challenges without exacerbating existing inequalities between developed and developing nations.

Chapter 9 then examines a critical but often overlooked consequence of environmental improvement — the phenomenon of "green gentrification." This chapter directly builds on the environmental justice themes introduced earlier, showing how even well-intentioned cleanup efforts can perpetuate the cycles of inequality. Through case studies of places like Greenpoint, Portland, and Seoul, it reveals how environmental cleanup efforts often lead to the displacement of vulnerable communities — the very populations most affected by historical pollution. The chapter analyzes how property values rise following air quality improvements, pushing out long-term residents who endured decades of pollution. It particularly focuses on the redevelopment of former industrial zones, documenting both successful inclusive approaches and cautionary tales of displacement. It calls for developing nations to attend to "just transition" policies that ensure environmental improvements benefit existing communities rather than disempowering them.

Chapter 10 investigates an emerging and controversial "solution" to the urban air pollution problem introduced in prior chapters — the construction of entirely new "clean" cities. From Indonesia's planned capital Nusantara to Neom at the northern tip of the Red Sea, governments and developers are attempting to engineer their way out of air quality crises by building from scratch. The chapter reveals how these projects, while technologically impressive, raise troubling questions about environmental justice and urban abandonment. It explores whether these satellite cities represent genuine innovation or merely allow elites to escape the consequences of pollution while leaving existing urban centers to deteriorate. The chapter also examines how these ambitious construction projects threaten to displace Indigenous peoples who have inhabited these lands for centuries, such as the Balik people in Indonesia and the Huwaitat tribe in Saudi Arabia, raising critical questions about whose voices are silenced in the pursuit of "sustainable" urban development.

The book then transitions to exploring alternative approaches and frameworks for addressing air pollution. Chapter 11 explores the potential of integrating traditional or Indigenous knowledge systems into modern environmental management strategies to address air pollution in developing countries. It highlights specific examples such as Indigenous Australian fire management practices, the use of neem leaves in India for air purification, and the application of Feng Shui principles in urban planning, demonstrating how ancient wisdom can align with and enhance contemporary scientific approaches. The chapter discusses the challenges in implementing these integrated approaches, including the need for scientific validation and adaptation to modern urban scales. It emphasizes the value of this integration in developing more holistic, culturally appropriate, and sustainable solutions to air quality challenges while simultaneously preserving cultural heritage and biodiversity.

Chapter 12 recognizes the impact of local grassroots movements against air pollution in developing countries, tracing their evolution from early environmental activism to modern-day initiatives. It highlights notable examples such as the Chipko movement in India, citizen-led efforts in Mexico City, and recent digital campaigns in China and Thailand, demonstrating how these movements have raised awareness and influenced policy. The chapter discusses the role of social media and digital technologies in empowering these grassroots efforts, as well as their focus on intersecting social and economic issues.

Chapters 13 and 14 then explore how private sector engagement can be reimagined to address air pollution more effectively. Chapter 13 explores how corporate–civic partnerships are emerging as a crucial bridge between profit-driven enterprises and community needs in addressing air pollution. It begins by examining the limitations of both purely corporate-led initiatives (often prioritizing shareholder value) and isolated grassroots movements (sometimes lacking resources and scale), setting up the case for strategic collaboration. It addresses how these partnerships can help avoid the "race-to-the-bottom" phenomenon described in Chapter 4 and reconcile the gap between developed and developing nations' approaches to air quality management.

Chapter 14 builds upon the corporate–civic partnerships discussed in Chapter 13 to explore how environmental metrics are fundamentally reshaping business models and corporate definitions of success. It examines the evolution of Environmental, Social, and Governance (ESG) criteria, particularly focusing on air quality indicators and their integration into corporate decision-making. Through case studies of companies like Patagonia and Interface, the chapter demonstrates how businesses are moving beyond traditional metrics to embrace environmental stewardship as a core measure of success. It critically analyzes the rise of Benefit Corporations and related certification systems, examining their potential to create systemic change while acknowledging their limitations.

Finally, Chapter 15 charts the emergence of clean air as a fundamental human right through groundbreaking legal cases and constitutional reforms. It examines how citizens and advocacy groups are increasingly framing air pollution not just as an environmental challenge but as a human rights violation, leading to novel forms of litigation and activism. The chapter explores precedent-setting cases where courts have recognized the right to breathe clean air and the implications of treating clean air as a human right rather than a commodity. This rights-based framework offers new tools for addressing global air pollution while challenging traditional approaches to environmental protection.

This work aims to provide a comprehensive understanding of the global air pollution crisis, its causes, and the innovative solutions emerging worldwide. It offers a unique perspective on how human nature, economic aspirations, and environmental concerns intersect, providing valuable insights into the complex challenges faced by both developed and developing nations. Lastly, it seeks to illuminate the intricate interplay between policy, technology, corporate decision-making, and social action in addressing air pollution.

Mass Production, Mass Consumption, and the Vanishing Blue

Chapter 1

From Factories to Firmament: Offshoring Industries Across Borders

Once upon a time, the skies of the Western world were shrouded in a perpetual haze. Factory smokestacks belched thick plumes of acrid smoke into the air, coating cities in a grimy film and turning the day into a murky twilight. This was the world Charles Dickens vividly portrayed in his novels, a world where industrial progress came at a steep cost to human health and the environment. In his book *Hard Times*, Dickens (1854) described the fictional industrial setting of Coketown (p. 32):

> It was a town of red brick, or of brick that would have been red if the smoke and ashes had allowed it; but as matters stood, it was a town of unnatural red and black like the painted face of a savage. It was a town of machinery and tall chimneys, out of which interminable serpents of smoke trailed themselves for ever and ever, and never got uncoiled.

This vivid imagery captured the reality of many industrialized cities in the 19th and early 20th centuries. London's infamous "pea-souper" fogs, a toxic mix of smoke and fog, claimed thousands of lives. During the period from Friday, December 5th to Tuesday, December 9th in 1952, visibility in many parts of England's capital city dropped so low that workers got lost in familiar streets and alleys. People were forced to abandon their vehicles and left gasping for air, and in many places, ranchers found their cattle asphyxiated. A letter from medical practitioner L. F. Beccle to the

Ministry of Health written in December 1952 noted the following (Robson-Mainwaring, 2022):

> I would have shared the fate of the Aberdeen Angus cattle at the Smithfield Show, for whom I had great sympathy and fellow feeling. I could not move for four days without the greatest distress ...

Thousands of Londoners found themselves living through a nightmarish five days where each labored breath could be their last — an experience so traumatic it would leave an indelible mark on the collective memory of a generation that had survived war only to face death in the very air they inhaled.

According to Polivka (2018), the Great Smog of 1952 killed an estimated 4,000–12,000 people and ultimately prompted the passage of the Clean Air Act of 1956 in the United Kingdom. The mortality numbers reported initially by the officials at the time of the incident were later reassessed and found to be much higher (Bell *et al.*, 2008; Logan, 1953). In subsequent years, when elevated rates of mortality persisted, more research was conducted on the mechanisms linking air pollution to altered death rates (Bell and Davis, 2001).

Similar stories unfolded across Europe and North America as the Industrial Revolution progressed. For example, back in December 1930, smog in Meuse Valley, Belgium, a leading steel industry site, claimed the lives of 63 individuals. There were also hundreds of cases of acute pulmonary attacks in Meuse Valley between the 1st and 5th of December, 1930. A public inquiry, which involved 15 autopsies, concluded the cause lay in poisonous substances in the smoke released by the many factories in the valley, in conjunction with unusual meteorological conditions (Firket, 1936). Nemery *et al.* (2001) observed that this incident was the "first scientific proof of the potential for atmospheric pollution to cause deaths and disease" (p. 704).

Likewise, in the United States, the deadly Donora Smog in Pennsylvania resulted in the deaths of 20 people and sickened thousands of others in 1948 (Jacobs *et al.*, 2018). The event helped to trigger the clean air movement in the United States and was pivotal in spurring Congress to pass the Clean Air Act in 1963, with President Nixon subsequently creating the Environmental Protection Agency (EPA) in 1970. The organization's initial purpose was to fix ineffective federal regulations to control pollution and to monitor and enforce them.

As environmental regulations tightened, industries in developed countries like the United Kingdom and the United States faced increasing costs to comply with new standards. These expenses included the installation of pollution control equipment, upgrades to more efficient and cleaner technologies, waste management and disposal, environmental impact assessments and monitoring, and potential fines and legal costs for non-compliance. Surveys by the EPA found that pollution abatement capital expenditures in manufacturing industries in the United States increased by several billion dollars from 1973 to 1993, a significant financial burden for many companies (EPA, 1993). Faced with rising costs and stringent regulations, these firms began to look for cost-cutting alternatives in places that would allow them to mass manufacture products without having to comply with strict environmental guidelines. Developing countries, eager for economic growth and foreign investment, presented an attractive option. This led to a significant shift of manufacturing and heavy industries from developed to developing nations, particularly from the 1980s onwards.

An internal World Bank memo from 1991 by Lawrence H. Summers stated, "I think the economic logic behind dumping a load of toxic waste in the lowest wage country is impeccable and we should face up to that .... I've always thought that underpopulated countries in Africa are vastly under polluted; their air quality is vastly inefficiently low compared to Los Angeles or Mexico City" (Mokhiber and Weissman, 1997). While Summers clarified that his remarks were intended as irony, the memo's cold economic rationale — suggesting that toxic waste should be dumped in lowest-wage countries — starkly illuminates how economic imperatives can be used to justify profound environmental injustice. When pollution occurs in distant places, affecting people who seem remote and different, it becomes easier for institutions and individuals to rationalize these choices through purely economic metrics. Complex bureaucracies create layers of distance between decisions and their consequences, while technical economic language masks the fundamental ethical implications. The focus on financial optimization leads to a system of modern continuation of colonial exploitation, where environmental costs are systematically externalized by powerful corporations to those with the least power to resist.

Looking back at history, there were certainly many prepared to sacrifice people elsewhere to protect their own backyard. Over time, numerous "dirty" industries, typically characterized by their significant negative impact on the environment through high levels of pollution, resource depletion, or both, relocated from developed to developing countries. The textile industry, known for its heavy use of water and chemicals, was one of the first to shift operations to developing countries. In the 1980s and 1990s, many textile manufacturers moved from the United States and Europe to countries like China, Bangladesh, and Vietnam. Between 1973 and the present, thousands of textile and apparel manufacturing jobs were lost in the United States as production shifted overseas. Similarly, the steel industry, another significant source of air pollution, also saw a major shift from the US and Europe to countries like China, India, and Vietnam. Steel mills in these destination countries often operated without basic pollution controls that would be mandatory in the US or the European Union, leading to severe local air quality impacts and higher global emissions per ton of steel produced. This dramatic shift not only changed the economic landscape but also transferred the environmental burden of steel production to the developing world. Relatedly, the electronics industry, which involves the use of various chemicals and generates significant e-waste, largely moved to Asia. Foxconn, a major manufacturer for companies like Apple, employed over a million workers in China by 2012, highlighting the massive scale of this industrial migration.

Political will, trade agreements, and regulatory actions often facilitated these transitions. Following the North American Free Trade Agreement's (NAFTA) implementation in 1994, there was a documented surge in pollution-intensive industries moving to Mexico, particularly in the maquiladora border region. Industries with high environmental compliance costs in the US — including electronics, chemical manufacturing, and automotive parts — increasingly shifted production to Mexico. The border region saw a concentration of hazardous waste sites and air pollution hotspots. Likewise, in East Asia, after Japan implemented strict environmental regulations in the 1970s, major corporations like Asahi Glass and Nippon Chemical began relocating their pollution-intensive operations to Southeast Asia. This shift was particularly pronounced in "dirty" industries like petrochemicals, plastics, and heavy metals processing. In Thailand and Indonesia, these companies established manufacturing facilities in areas with minimal environmental oversight.

Here, it is important to note that the designation of an industry as "dirty" is not static. Technological advancements and stricter regulations can transform once-dirty industries. For instance, the automotive industry, once a significant polluter, has made major strides in reducing emissions through electric and hybrid technologies. Even for conventional vehicles, advancements in engine technology have dramatically improved fuel efficiency and reduced emissions. Legacy automakers like Volvo have committed to becoming a fully electric car company by 2030 and to be climate neutral by 2040, although a recent decline in sales has led its leadership to reconsider how much time it would take for full electrification.

Back in the 1980s and 1990s, developing countries were often eager to attract these heavily polluting industries for several reasons. Industrial development was seen as a path to rapid economic growth and employment generation. For instance, China's economic reforms starting in 1978 prioritized industrialization, leading to a relatively high annual GDP growth for nearly three decades. Moreover, these countries also hoped to gain access to advanced technologies through foreign investments. To illustrate, India's auto industry saw significant technological upgrades following economic liberalization in 1991. In addition, manufacturing industries provided valuable export earnings for developing economies. Bangladesh's ready-made garment exports, for instance, grew from mere millions in 1983 to a multi-billion-dollar industry in 2012–2013, transforming the country's economy in the process.

Many developing nations viewed industrialization as a necessary step in the development process, following the path taken by Western nations. This view was influenced by Western-centric development theories like Rostow's stages of economic growth. Environmental issues were often seen as a luxury that could be addressed after achieving a certain level of economic progress. This perspective was famously articulated by former Indian Prime Minister Indira Gandhi at the *United Nations Conference on Human Environment* in Stockholm on 14th June, 1972: "Aren't poverty and need the greatest polluters?"

As industries migrated, so did air pollution. This created what Rob Nixon (2011) calls "slow violence" — a gradual and often invisible form of environmental degradation that disproportionately affected the poor. While air quality in many Western cities improved, pollution levels in developing countries' urban areas skyrocketed. In the United States, the average concentration of PM2.5 decreased by 37% between 1990 and 2015 (NASA, n.d.). In contrast, PM2.5 concentrations in many cities in East Asia and South Asia increased dramatically during the same period. Relatedly, Wang *et al.* (2017) demonstrated that during the 1990–2010 period, PM2.5 mortalities increased by 21% and 85% in East Asia and South Asia, respectively, whereas in Europe and North America, they decreased substantially by 67% and 58%, respectively.

The environmental impact of industrial migration was further amplified by the rise of mass production and mass consumption. Offshoring allowed for larger-scale production facilities, often with lower per-unit costs but higher total emissions. For example, China's Tangshan region became the world's largest steel-producing area, accounting for 8% of global output by 2014 but also contributing significantly to regional air pollution (CRU, 2021). In 2023, the local government in Tangshan asked 11 Class A steel mills in the region to take the lead in reducing production by about 30% in an effort to curb air pollution.

Similarly, the textile industry's shift to developing countries coincided with the rise of "fast fashion," characterized by rapid production cycles and low-cost, disposable clothing. This led to increased production volumes and associated pollution. Global clothing production doubled between 2000 and 2014, putting immense pressure on the environment in producing countries like Bangladesh. The electronics industry also illustrates this trend. As demand for consumer electronics boomed in the West, much of the production — and associated pollution — shifted to countries like China. Shenzhen, once a small fishing village, transformed into a major electronics manufacturing hub, earning the nickname "the world's factory" (Al, 2015).

The offshoring of pollution-intensive industries from developed nations to developing countries has had far-reaching consequences. On the one hand, in countries like the United States and the United Kingdom, the impact of deindustrialization went beyond mere employment and air quality statistics, cutting deep into the psychological and emotional well-being of affected workers and their families. Many workers experienced not just the loss of income but a profound crisis of identity. Manufacturing jobs had provided not merely employment but a sense of purpose, pride, and place within society. The loss of these positions struck at the heart of how many workers, particularly men, defined themselves and their roles in their families and communities.

The impact on communities proved equally profound. Towns and cities built around manufacturing faced not just economic decline but a dissolution of their very reason for existence. Places like Youngstown, Ohio, and Sheffield, England, experienced not only population decline and property value collapse but a deeper crisis of communal purpose (Rhodes, 2013). Social institutions that had evolved around manufacturing — unions, clubs, church groups, and neighborhood associations — began to unravel as their membership dispersed. Traditional values of hard work, loyalty, and craftsmanship seemed increasingly at odds with a new economic reality that prioritized flexibility, service, and digital skills. Russo and Linkon (2009) observed that over the three decades following deindustrialization, cities like Youngstown saw "declines in population and tax base, and battled persistently high crime rates, urban decay and questions about whether there just wasn't something wrong with the community" (p. 149).

On the other hand, offshoring pollution-intensive industries created global inequality in environmental quality, with citizens in developing countries bearing a disproportionate burden of pollution. In particular, the health impacts of air pollution have been severe, including causing significant limitations to growth, as well as lung function deficits, among children. Odo *et al.* (2022) found that annual ambient PM2.5, as an indicator of long-term exposure, was associated with greater odds of maternal-reported acute respiratory infections in children aged less than five years living in 35 low- and middle-income countries. Much of the world's population now lives in places where air quality guidelines are not met, with low- and middle-income countries suffering from the highest exposures. Thus, while developing countries have benefited from industrial growth, they now face significant economic and health costs due to air pollution.

The concept of the "privilege of blue skies" encapsulates this disparity, where Western countries enjoy cleaner air and water while developing countries grapple with the environmental consequences of industrial production. People living in the former can breathe easier, live longer, and better enjoy nature's beauty. However, the "privilege of blue skies" extends beyond just cleaner air. Western consumers benefit from access to a wide range of affordable goods produced in developing countries, often without bearing the full environmental costs of this production. Meanwhile, workers and communities in developing countries face not only pollution but also related health issues.

In Franz Kafka's (1915/2013) *The Metamorphosis*, salesman Gregor Samsa awakens one morning to find himself transformed into a monstrous insect, a surreal change that serves as a powerful metaphor for alienation and powerlessness in modern society. Similarly, the global shift of industrial production from West to East represents its own kind of metamorphosis — one that has transformed both landscapes and lives in equally profound, if less visibly dramatic, ways. Just as Gregor's family benefits from his labor while recoiling from his transformed state, Western societies have undergone their own convenient transformation. They have shed their industrial "carapace," exporting it to distant shores while maintaining their consumption patterns and lifestyle benefits. The "privilege of blue skies" becomes their version of the clean, orderly apartment that Gregor's family maintains, while deliberately ignoring the creature that made their comfort possible.

The parallel becomes even more striking when we consider the psychological state of workers in developing nations' industrial zones. Like Gregor, they find themselves trapped in a physically degraded state, serving an economic system that views them as necessary but ultimately disposable components. They too experience a form of metamorphosis — their bodies and environments gradually transformed by exposure to industrial pollutants, their lives increasingly alienated from the natural world they once knew.

In Kafka's story, the family's gradual acceptance of Gregor's condition mirrors the way both Western consumers and developing world workers have normalized their respective positions in this global environmental divide. Western consumers, like Gregor's family, have learned to look away from the unsettling transformation occurring in distant lands. They navigate air-conditioned malls and scroll through online shopping platforms, their connection to the environmental cost of their consumption as remote as Gregor's family's understanding of his suffering. Meanwhile, the workers in developing nations' industrial zones, like Gregor himself, experience a profound disconnection from their environment. They inhabit spaces where the air burns their lungs and water runs with unnatural colors, yet they must continue their daily routines as if this degradation were normal. Like Gregor's inability to communicate his anguish in human language, affected communities often find their concerns dismissed or untranslatable in the language of economic priorities and policy.

In *The Metamorphosis*, Gregor's room becomes increasingly cluttered and dirty as the story progresses, while the rest of the apartment remains pristine. This spatial division eerily reflects our global environmental arrangement — Western nations enjoy their clean air and water, while developing regions become the world's industrial "room," accumulating the waste and pollution that makes this cleanliness possible. This Kafkaesque lens reveals how the "privilege of blue skies" is maintained through a system of deliberate disconnection and denial, where the transformation of some enables the comfortable stasis of others. It challenges us to question whether we, like Gregor's family, have become too accustomed to building our comfort upon others' metamorphosis into something less than what they once were.

The story of how the West exported its pollution to developing countries is a complex tale of economic incentives, regulatory pressures, and global power dynamics, which highlights the interconnected nature of our global economy and environment, where actions in one part of the world can have far-reaching consequences elsewhere. As we stand at this critical juncture, with pollution and environmental degradation posing existential threats, the lessons from this history are clear. We cannot solve our environmental problems by simply moving them elsewhere. True solutions will require global cooperation, innovative technologies, and a fundamental rethinking of our production and consumption patterns. The journey from Dickens's smoke-filled Coketown to a world where everyone can enjoy clean air is long and challenging, but it is a journey we must undertake for the sake of our planet and future generations.

References

Al, S. (2015, February 9). Mass-producing the world's factory. *Metropolitics.* Retrieved on February 23, 2024 from http://www.metropolitiques.eu/Mass-Producing-the-World-s-Factory.html.

Bell, M. L. and Davis, D. L. (2001). Reassessment of the lethal London fog of 1952: Novel indicators of acute and chronic consequences of acute exposure to air pollution. *Environmental Health Perspectives, 109*(suppl 3), 389–394.

Bell, M. L., Davis, D. L., and Fletcher, T. (2008). A retrospective assessment of mortality from the London smog episode of 1952: The role of influenza and pollution. In J. M. Marzluff *et al.* (Eds.) *Urban Ecology* (pp. 263–268). Boston, MA: Springer.

CRU. (2021, May 6). CRU explains: How Tangshan influences the global steel market. Retrieved on May 20, 2024 from https://www.crugroup.com/knowledge-and-insights?type=1601.

Dickens, C. (1854). *Hard Times: A Novel.* Franklin Square, NY: Harper & Brothers.

EPA. (1993). Previously Published PACE Survey Data (1973–1993). *United States Environmental Protection Agency.* Retrieved on May 20, 2024 from https://www.epa.gov/environmental-economics/previously-published-pace-survey-data-1973-1993.

Firket, J. (1936). Fog along the Meuse valley. *Transactions of the Faraday Society, 32*, 1192–1196.

Jacobs, E. T., Burgess, J. L., and Abbott, M. B. (2018). The Donora smog revisited: 70 years after the event that inspired the clean air act. *American Journal of Public Health, 108*(S2), S85–S88.

Kafka, F. (1915/2013). *The Metamorphosis* (S. Corngold, Trans.). New York, NY: Modern Library.

Logan, W. P. (1953). Mortality in the London fog incident, 1952. *The Lancet, 261*(6755), 336–338.

Mokhiber, R. and Weissman, R. (1997). Memo misfire: World Bank "spoof" memo on toxic waste holds more irony than laughs. *San Francisco Bay Guardian.* Retrieved on May 22, 2024 from https://archive.globalpolicy.org/socecon/bwi-wto/sumers99.htm.

NASA. (n.d.). U.S. air quality trends: Efficacy of environmental regulations to improve air quality in the U.S. *National Aeronautics and Space Administration.* Retrieved from https://airquality.gsfc.nasa.gov/us-air-quality-trends.

Nemery, B., Hoet, P. H., and Nemmar, A. (2001). The Meuse Valley fog of 1930: An air pollution disaster. *The Lancet, 357*(9257), 704–708.

Nixon, R. (2011). *Slow Violence and the Environmentalism of the Poor.* Cambridge: Harvard University Press.

Odo, D. B. *et al.* (2022). Ambient air pollution and acute respiratory infection in children aged under 5 years living in 35 developing countries. *Environment International, 159,* 107019. doi: https://doi.org/10.1016/j.envint.2021.107019.

Polivka, B. J. (2018). The great London smog of 1952. *American Journal of Nursing, 118*(4), 57–61.

Rhodes, J. (2013). Youngstown's 'ghost'? Memory, identity, and deindustrialization. *International Labor and Working-Class History, 84,* 55–77.

Robson-Mainwaring, L. (2022, July 19). The Great Smog of 1952. *The National Archives.* Retrieved on May 22, 2024 from https://blog.nationalarchives.gov.uk/the-great-smog-of-1952/.

Russo, J. and Linkon, S. L. (2009). The social costs of deindustrialization. In R. McCormack (Ed.), *Manufacturing a Better Future for America* (pp. 183–216). Washington D.C.: Alliance for American Manufacturing.

Wang, J., Xing, J., Mathur, R., Pleim, J. E., Wang, S., Hogrefe, C., Gan, C., Wong, D., and Hao, J. (2017). Historical trends in PM2. 5-related premature mortality during 1990–2010 across the northern hemisphere. *Environmental Health Perspectives, 125*(3), 400–408.

Chapter 2

Clouded Rainbows: Pricing Consumerist Aspirations and Quality-of-Life Pursuits

There had been a time when our ancestors lived in a world where the concept of "disposable" was nearly nonexistent. They embraced a lifestyle of simplicity, frugality, and resourcefulness that stands in stark contrast to our modern "use-and-throw" culture. This shift from contentment with less to an insatiable appetite for more is a tale of societal transformation, economic growth, and unforeseen consequences. In pre-industrial societies, people valued durability and repairability. Clothes were mended, tools were crafted to last generations, and household items were cherished possessions. Traditional Japanese culture exemplified this philosophy through the word "mottainai" — a term expressing regret over waste and a deep respect for the intrinsic value of objects (Taylor, 2023). Similarly, Indigenous peoples in America practiced principles of sustainable consumption long before it became a modern environmental concern, using every part of hunted animals and maintaining a spiritual connection with their material possessions. As Henry Thoreau (1854/2006) observed in *Walden*, "Our life is frittered away by detail ... Simplicity, simplicity, simplicity!" This ethos of simplicity was not just a philosophical stance but a practical necessity in a world of limited resources and production capabilities.

The Industrial Revolution marked a turning point in human history, fundamentally transforming not just production methods but the very relationship between humans and material goods. This transformation was starkly illustrated in Manchester, England, where the introduction of steam-powered machinery revolutionized textile production, radically intensifying cotton processing between the 1780s and 1850s. Yet this shift from artisanal to mass production came with profound social costs that would foreshadow many of the contradictions in modern consumer society. Workers who once took pride in crafting entire garments now performed repetitive tasks in deafening, dangerous factories before returning to cramped slums with inadequate sanitation. The combination of coal-burning factories and inadequate ventilation created thick smog that directly impacted workers' health. Friedrich Engels (1844) described the atmosphere being "enveloped in a grey cloud of coal smoke" (p. 53).

The human toll was severe — in 1819, about 41% of children born to workers at McConnel and Kennedy mills did not survive to adulthood (Stallard *et al.*, 2023), highlighting how the drive for efficient production often sacrificed human welfare. The influx of Irish migrants into the lowest-paid factory work created new patterns of social stratification based on consumption ability, a precursor to modern consumer-based class distinctions. Even as factory owners accumulated unprecedented wealth and their families enjoyed new levels of material comfort, working-class families struggled to afford basic necessities, creating stark disparities in consumption patterns that would become characteristic of industrial societies. The Peterloo Massacre of 1819, where peaceful protesters demanding better conditions were violently suppressed (Poole, 2019), illustrated the growing tensions between production, consumption, and social justice — tensions that would continue to shape consumer society for generations to come.

As countries industrialized, there was indeed some correlation between economic growth and enhanced living standards, with improvements in clean water, sanitation, and access to education becoming more widespread. However, these developments also catalyzed a fundamental shift in social values and consumer behavior. The ability to mass-produce goods at increasingly lower costs created new forms of social aspiration and status marking through material possessions — a pattern that would intensify over the following centuries. Writing as early as 1893, Mark Twain in his short story *The Million Pound Bank Note* offered a prescient critique of how material wealth and the appearance of consumption ability had become the primary determinants of social status. When the protagonist in Twain's story, Henry Adams, an impoverished clerk, possesses the million-pound note, he receives credit, respect, and opportunities simply because he appears wealthy — despite never actually spending the note.

This story brilliantly captured the emerging reality where the capacity for consumption, rather than character or craft, increasingly defined one's place in society.

The aftermath of World War I marked a profound transformation in Western society, as the trauma of unprecedented casualties collided with an emerging culture of consumerism and hedonistic pursuit. This period witnessed the rise of mass production techniques, expanding middle-class disposable income, and new advertising strategies that promised fulfillment through material consumption. Ernest Hemingway's (1926) *The Sun Also Rises* masterfully captures this cultural shift through its portrayal of expatriate Americans in post-war Europe. The novel's characters embody the desperate pursuit of pleasure and meaning in a world where traditional values have been shattered. Jake Barnes and his companions moved through an endless cycle of drinking, dancing, and romantic entanglements, their consumption of experiences and goods serving as an escape and as a means of outrunning existential despair. Their traveling from Paris cafés to Spanish fiestas mirrored the restless consumption patterns emerging in post-war society, where people increasingly sought meaning through purchasing power and leisure rather than traditional social bonds or spiritual practices.

The widespread adoption of private automobiles, encouraged by suburban development and highway construction, introduced a new source of persistent air pollution. Los Angeles, which embraced car culture earlier and more completely than other American cities, began experiencing severe smog episodes by the late 1940s. By 1943, the situation had become so severe that residents mistook the first major smog event for a Japanese chemical attack. This marked the beginning of a new era where personal consumption choices — specifically, the individual ownership of automobiles — would directly contribute to collective environmental degradation.

The subsequent World War II dramatically accelerated and intensified the consumerist trends that emerged after World War I, fundamentally reshaping American society and global consumption patterns. As factories retooled for civilian production after the war, this massive industrial infrastructure needed new markets. Simultaneously, Americans had accumulated significant savings during wartime rationing, while the GI Bill created a new middle class with purchasing power. Lebow (1955) observed, "Our enormously productive economy ... demands that we make consumption our way of life, that we convert the buying and use of goods into rituals, that we seek our spiritual satisfaction, our ego satisfaction, in consumption" (p. 7).

This transformation was not accidental but deliberately engineered. Post-World War II America saw a concerted effort by businesses and the government to stimulate consumption as a means of economic growth. President Eisenhower's Council of Economic Advisers Chairman Arthur Burns famously declared in 1953, "The American economy's ultimate purpose is to produce more consumer goods." Government policies actively encouraged consumption through programs like suburban housing developments and highway construction. The nuclear family in their single-family home, surrounded by modern appliances and automobiles, became both an American ideal and a Cold War weapon, demonstrating capitalism's superiority over socialism.

As the consumerist ideology spread globally, there rose an optimistic belief that mass production would eventually deliver material abundance to all. This optimism manifested differently across developed and developing nations but shared a common thread — the aspiration for a better life through increased consumption. In China, the transformation was particularly dramatic. During the 1970s, basic consumer goods in China were strictly rationed through a system of tickets (Chinn, 1980). Bicycles, considered a luxury item, required special purchase certificates, and their acquisition often represented a major family milestone. Meat consumption was so limited that many families could only afford it during major festivals. The reforms of the 1980s and 1990s dramatically changed this landscape (Chao and Myers, 1998). By the 1990s, bicycles became widely available, and by the 2000s, China had transformed into the world's largest automobile market. Meat consumption per capita multiplied several times between 1975 and 2015.

In India, similar patterns emerged, though with different timing and manifestations. In the 1970s, the waiting period for a Bajaj scooter could stretch to several years (Siddiqui, 1991), and items like refrigerators were considered luxury goods owned by only a small urban elite. The Green Revolution of the 1960s and 1970s had increased food production, but distribution remained a challenge. The economic liberalization of the 1990s transformed consumption patterns. The emergence of shopping malls, supermarkets, and consumer financing options created new possibilities for middle-class consumption. The humble scooter gave way to motorcycles and small cars, while household appliances became increasingly common in urban homes.

Likewise, in Southeast Asia, countries like Malaysia and Indonesia experienced analogous transformations. Malaysia's New Economic Policy (1971–1990) explicitly aimed to create a consumer middle class, promoting homeownership and automobile purchases. Indonesia's development under President Suharto, despite some problems, saw rising incomes translate into increased consumption of manufactured goods and processed foods. This optimism about consumption wasn't limited to physical goods. Education, healthcare, and leisure activities became part of the aspiration package. Shopping malls emerged as temples of consumption — from Bangkok's air-conditioned megamalls to Beijing's bustling markets. These spaces represented not just shopping venues but symbols of modernity and progress. They reflected not just material desires but deeply held aspirations for dignity, choice, and improved life circumstances — aspirations that continue to shape global economic and social development. For millions of parents, the ability to provide better education and healthcare for one's children through increased consumption capacity often generates genuine and lasting satisfaction.

However, the consumerist ideology also brought with it unforeseen costs. It often created temporary satisfaction followed by new desires. To illustrate, in the 1980s, just when some families in China felt that buying a Phoenix bicycle would make them satisfied and happy for a long time, their neighbors upgraded to motorcycles, making the former feel left behind with an urge to catch up. This aspirational stress to continuously play "catch up" led to competitive consumption, a phenomenon where individuals consume not merely for utility but to maintain social status within an ever-escalating hierarchy of material possessions. Over time, consumption became deeply intertwined with identity formation, and with the advent of social media that further highlighted relative deprivations, created constant pressure to maintain and update one's consumer identity. In China, consumption ability became a common metric for judging potential marriage partners (Grier *et al.*, 2016). Similar patterns emerged in India, where consumer goods became crucial markers of family status.

Industries responded to this psychology-driven demand by increasing production and introducing planned obsolescence. The fashion industry exemplifies this pattern — fast fashion companies moved from two seasons per year to as many as 52 micro-seasons, dramatically increasing production and environmental impact. Companies increasingly targeted psychological vulnerabilities, with terms like "retail therapy" and "comfort shopping" becoming normalized, while advertising explicitly played on social anxiety and status concerns. Kim and Chang (2023) detailed how Korean consumers exhibited an increase in revenge consumption of luxury products as a response to the COVID-19 pandemic, deemed one of their most stressful life events. In addition, Langefels (2023) explained how millennial female consumers often looked for affordable or bargain online shopping deals to cope with negative emotions and loneliness experienced during the pandemic. For many consumers, the joy of frequent purchases at affordable prices overshadowed other considerations.

The race to produce ever-cheaper goods created economic distortions globally. As mass production shifted to developing countries in search of cheaper labor, it created new pollution hotspots. Moran and Kanemoto (2016) describe the case of the Port of Los Angeles, which made significant progress in reducing pollution from trucks, incoming ships, and on-site sources, but at the same time, opened a new shipping terminal to increase imports from China, where air pollution had become a major problem. This scenario encapsulates the essence of what economist William Nordhaus (1999) terms the "global public goods problem" in environmental protection, where improvements in one area can be offset by deterioration elsewhere.

Similar patterns have been observed in other developed nations. For instance, the United Kingdom has seen a significant reduction in its territorial carbon emissions since 1990, largely due to the offshoring of manufacturing. However, when accounting for consumption-based emissions, which include those embedded in imported goods, the UK's carbon footprint has actually increased. This discrepancy highlights the limitations of national-level environmental policies in an interconnected global economy.

The driving force behind this global redistribution of pollution is fundamentally rooted in human nature, particularly the trait of self-interest. As philosopher Thomas Hobbes posited in his seminal work *Leviathan*, human beings are inherently self-interested, driven by a desire for self-preservation and betterment. In developed countries, this self-interest manifests in the desire to maintain a high quality of life afforded by mass consumption. The ability to enjoy clean air while pollution-intensive industries are offshored creates a psychological distance from the problem, allowing for a form of environmental cognitive dissonance. This is further exacerbated by what psychologists term the "construal level theory" (Trope and Liberman, 2010), which suggests that people tend to think about psychologically distant events in abstract terms, while near events are construed more concretely. Applied to environmental issues, this theory helps explain why some individuals in developed countries might struggle to connect their consumption habits with pollution in distant lands (Brügger *et al.*, 2016).

Conversely, in developing countries like China and India, self-interest takes on a different form. The promise of economic growth and poverty alleviation through industrialization presents an alluring path. Between 1978 and 2013, China's economy grew at an average annual rate of 9.5%, lifting over 800 million people out of poverty (Hu and Khan, 1997). However, this economic miracle came at a severe environmental cost. A study by Liu *et al.* (2017) estimated that "national PM2.5 related deaths from stroke, ischemic heart disease and lung cancer increased from approximately 800,000 cases in 2004 to over 1.2 million cases in 2012" (p. 75). This prioritization of immediate economic benefits over long-term environmental costs reflects what behavioral economists call "hyperbolic discounting," a tendency for people to choose smaller, immediate rewards over larger, delayed rewards (Laibson, 1997). In the context of environmental policy, this human tendency often leads to the postponement of crucial but costly environmental measures in favor of short-term economic gains.

However, while psychological distance might obscure the immediate perception of environmental damage in developed nations and economic pressures drive environmental degradation in developing ones, the physical manifestations of these combined effects have become increasingly difficult to ignore. The original post-war vision had not adequately anticipated how mass consumption would interact with environmental sustainability issues. A stark example of this oversight can be found in the Great Pacific Garbage Patch, also known as the Pacific trash vortex, brought to attention in 1997 by oceanographer Charles Moore. This massive concentration of plastic debris, now covering an area roughly twice the size of Texas, serves as a haunting monument to our throwaway culture. The phenomenon of "fast furniture" (Leroux *et al.*, 2023), popularized by companies like IKEA, has created a new category of environmental waste, with millions of tons of furniture ending up in landfills annually in the United States alone. Perhaps most tragically, many of us have become trapped in patterns of consumption that we know are unsustainable but feel powerless to change. This is further complicated by consumption lock-ins, where infrastructural, institutional, and social factors make it increasingly difficult for individuals to opt out of high-consumption lifestyles, even when they actively wish to do so.

The challenge, then, lies in reconciling these competing interests — the desire for economic development and poverty alleviation in developing countries, and the maintenance of high living standards in developed nations — with the imperative of global environmental protection. This tension between awareness and action, between individual desire for change and systemic constraints, points to the need for fundamental reforms in how we conceptualize self-interest and collective welfare. Economist Amartya Sen argued for an expanded notion of self-interest that includes concern for others and for future generations. This perspective aligns with recent developments in evolutionary biology and psychology, which suggest that human beings are capable of both selfish and cooperative behaviors and that our definition of "self" can expand to include larger groups, even humanity as a whole. In addition, the concept of green growth, championed by institutions like the Organization for Economic Co-operation and Development (OECD), posits that economic growth can be achieved while simultaneously reducing environmental impacts (Ates and Derinkuyu, 2021). This approach seeks to align self-interest with environmental protection by demonstrating the economic benefits of sustainable practices.

The trajectory of modern consumerism represents a profound transformation from its hopeful post-war origins to today's complex crisis of overconsumption. In the aftermath of World War II, the promotion of consumption emerged from genuinely optimistic intentions — a vision of creating widespread prosperity and preventing the economic conditions that had contributed to global conflict. Mass production and consumption promised to lift millions out of poverty, providing decent housing, adequate food, and basic comforts that were once luxuries. This perspective wasn't merely economic; it carried moral weight. The ability of ordinary families to own refrigerators, washing machines, and automobiles represented not just material progress but also human dignity.

However, this optimistic beginning gradually transformed as deeper human psychological vulnerabilities intersected with expanding productive capabilities and corporate profit maximization motives. Many societies jumped from scarcity to excessive consumption without experiencing the intermediate stage of sustainable consumption. This rapid transition created new social pressures and environmental strains while disrupting traditional values and community bonds. Social media transformed normal human desires for status and belonging into constant anxiety about consumption and appearance. This digital acceleration of consumption pressures created a perpetual inadequacy syndrome — a constant feeling of falling behind despite increasing consumption.

The journey from Manchester's coal-darkened skies to China's air pollution crisis, from simple tools cherished for generations to mountains of fast furniture in landfills, tells a story of progress twisted by psychological vulnerabilities and corporate interests. While mass consumption initially promised dignity and comfort for all, its current manifestation threatens not just environmental sustainability but also human well-being, creating a peculiar paradox where increased material abundance might yield diminishing returns in happiness. The challenge ahead lies not in abandoning consumption entirely, but in reimagining it in a way that honors both human needs and planetary boundaries — perhaps by rediscovering some of the wisdom embedded in traditional philosophies like "mottainai" while embracing innovative sustainable technologies and social systems. This transformation will require more than individual choice or technological solutions alone; it demands a fundamental restructuring of how we derive meaning, status, and satisfaction in our lives, moving from a paradigm of endless accumulation to one of mindful sufficiency.

References

Ates, S. A. and Derinkuyu, K. (2021). Green growth and OECD countries: Measurement of country performances through distance-based analysis (DBA). *Environment, Development and Sustainability*, *23*(10), 15062–15073.

Brügger, A., Morton, T. A., and Dessai, S. (2016). "Proximising" climate change reconsidered: A construal level theory perspective. *Journal of environmental psychology*, *46*, 125–142.

Chao, L. and Myers, R. H. (1998). China's consumer revolution: The 1990s and beyond. *Journal of Contemporary China*, *7*(18), 351–368.

Chinn, D. L. (1980). Basic commodity distribution in the People's Republic of China. *The China Quarterly*, *84*, 744–754.

Engels, F. (1844/1993). *The Condition of the Working Class in England*. Oxford, UK: Oxford University Press.

Grier, K. B., Hicks, D. L., and Yuan, W. (2016). Marriage market matching and conspicuous consumption in China. *Economic Inquiry*, *54*(2), 1251–1262.

Hemmingway, E. (1926). *The Sun also Rises*. New York: Modern Library.

Hu, Z. and Khan, M. (1997). Why Is China Growing So Fast? *International Monetary Fund, Economic Issues No. 8*. Washington D.C.: IMF Publication Services.

Kim, S. and Chang, H. J. J. (2023). Mechanism of retail therapy during stressful life events: The psychological compensation of revenge consumption toward luxury brands. *Journal of Retailing and Consumer Services, 75*, 103508. doi: https://doi.org/10.1016/j.jretconser.2023.103508.

Laibson, D. (1997). Golden eggs and hyperbolic discounting. *The Quarterly Journal of Economics, 112*(2), 443–478.

Langefels, E. K. (2023). *Coping with the Pandemic: A Qualitative Exploration of How Female Millennial Consumers Use Retail Therapy*. Doctoral dissertation, University of Minnesota.

Lebow, V. (1955). Price competition in 1955. *Journal of Retailing, 31*(1), 5–10.

Leroux, N., Kruckenberg, S., Moeller, E., Schwendemann, M., and Sigmund, M. (2023). Can Fast Furniture Really Be Sustainable? — An Analysis of the IKEA Business Model. *Management Studies, 13*(2), 3–16.

Liu, M., Huang, Y., Ma, Z., Jin, Z., Liu, X., Wang, H., Liu, H., Wang, H., Liu, Y., Wang, J., Jantunen, M., Bi, J., and Kinney, P. L. (2017). Spatial and temporal trends in the mortality burden of air pollution in China: 2004–2012. *Environment International, 98*, 75–81.

Moran, D. and Kanemoto, K. (2016). Tracing global supply chains to air pollution hotspots. *Environmental Research Letters, 11*(9), 094017. doi: 10.1088/1748-9326/11/9/094017.

Nordhaus, W. D. (1999, June). Global public goods and the problem of global warming. In *Annual Lecture of the 3rd Toulouse Conference of Environment and Resource Economics, Toulouse* (pp. 14–16).

Poole, R. (2019). *Peterloo: The English Uprising*. Oxford, UK: Oxford University Press.

Siddiqui, F. U. (1991). *Growth pattern of automobile industry in India since 1970*. Dissertation thesis: Aligarh Muslim University.

Stallard, M., Blood, D., and McMullan, L. (2023, April 3). How slavery made Manchester the world's first industrial city. *The Guardian*. Retrieved on January 4, 2024 from https://www.theguardian.com/news/ng-interactive/2023/apr/03/cotton-capital-how-slavery-made-manchester-the-worlds-first-industrial-city.

Taylor, K. (2023, March 24). Mottainai: A Japanese philosophy of waste. *Japan Up Close*. Retrieved on January 4, 2024 from https://japanupclose.web-japan.org/techculture/c20230324_3.html.

Thoreau, H. D. (1854/2006). *Walden*. New Haven, CT: Yale University Press.

Trope, Y. and Liberman, N. (2010). Construal-level theory of psychological distance. *Psychological Review, 117*(2), 440–463.

Chapter 3

Pricing the Last Breath: The Global Hierarchy of Human Worth

On a quiet summer evening on August 10th 1978, three young women — sisters Judy and Lynn Ulrich, 18 and 16 years old, respectively, and their cousin Donna, 18 — set out for a church-sponsored volleyball game. After a routine stop to fill their 1973 Ford Pinto's gas tank, the trio continued their drive on US-33 in northern Indiana, a road that unexpectedly would not lead to their intended destination. Around 6:30 p.m., as shadows lengthened across their path, Judy activated her hazard lights and began to slow down, but the driver of a Chevrolet behind them, momentarily distracted, failed to notice (Hartz, 2020). What began as a seemingly minor rear-end collision transformed into a nightmare when the Ford Pinto's flawed fuel tank ruptured, engulfing the vehicle in an inferno and trapping the teens inside. Three beautiful souls, who had woken that morning to the promise of tomorrow, would never again feel the summer breeze or their mothers' embrace. Their families were left with three flower-laden caskets, three empty bedrooms, and the endless echo of silence where young voices should have been — a silence that screamed of lives unlived and futures forever stolen. The tragedy sparked one of the most significant product liability cases in American history, leading to Ford's criminal prosecution. The subsequent trial would forever change how automotive safety was approached in the United States, though the verdict itself proved controversial when Ford was ultimately acquitted.

About a year before this grievous accident, in the fall of 1977, Mark Dowie's explosive exposé in *Mother Jones* had revealed how Ford Motor

Company estimated that it would be cheaper to pay for wrongful death lawsuits than to fix a deadly design flaw in their Pinto model. The company's internal memo showed a cold calculation: At $200,000 per death (about $1 million in today's terms), paying for an estimated 180 deaths would cost less than recalling and modifying the cars at $11 per vehicle. Dowie (1977) noted that Ford's memo clearly argued that there was "no financial benefit in complying with proposed safety standards that would admittedly result in fewer auto fires, fewer burn deaths and fewer burn injuries." Years later, this infamous case became a textbook example of how human life is assigned a monetary value in corporate and policy decisions. What many of us may not realize is that this practice continues today on a global scale, with even more striking disparities between nations.

The Value of Statistical Life (VSL) — the amount society is willing to pay to save one statistical life — has evolved significantly in developed nations. In the United States, the Environmental Protection Agency (EPA, n.d.) currently uses a VSL of $7.4 million (in 2006 dollars). This figure, derived from wage premium studies and stated preference surveys, represents how much Americans collectively value reducing mortality risks. In other words, Americans are willing to give up about $74 to reduce a 1-in-100,000 risk of death. The dramatic increase in VSL from the 1970s to 2006 reflects both economic growth and shifting societal values about human life. The EPA and other agencies use VSL figures in cost–benefit analyses of regulations and policies that affect public health and safety. For example, this number was used to justify the implementation of the Clean Air Act in the United States from 1990 to 2020 (Becerra-Pérez et al., 2024). It was also used during the COVID-19 pandemic to estimate the costs of continuing severe restrictions in various countries (Miles et al., 2020).

However, this evolution in VSL hasn't been uniform across the globe. The World Bank and other international organizations often use dramatically lower VSLs for developing nations, typically adjusting them based on GDP per capita. This leads to stark disparities — while a statistical life in the United States is valued at $7.4 million, the implied value in many developing nations can be as low as $100,000–500,000 (Sweis, 2022). These differences create a troubling economic rationale for the migration of hazardous industries.

Consider two identical factories with identical pollution control upgrade costs of $5 million. In a developed nation, preventing one statistical death through these upgrades is "worth it" if using the US EPA's $7.4 million VSL. However, in a developing nation where the VSL might be $500,000, preventing even 10 statistical deaths wouldn't justify the same investment from a purely economic perspective. This cold calculus provides one perspective on why pollution-intensive industries often relocate to regions with lower VSLs. This same calculus plays out in decisions about air quality controls and emissions standards. A factory in a developed nation might be required to install expensive scrubbers and filters to protect nearby communities, while its counterpart in a developing country might operate with minimal pollution controls, its toxic emissions drifting through neighborhoods where residents cannot afford to relocate. The mathematics of differential VSLs thus determines not just workplace safety but the very air communities breathe.

The pattern becomes even more stark when examining specific cases. When multinational corporations conduct cost–benefit analyses for safety and environmental investments, they often use different VSLs for different countries. This practice, while controversial, is defended as reflecting local economic conditions and willingness to pay as a society. However, it raises profound ethical questions: Should the worthiness of saving a human life vary based on geography? Does economic pragmatism justify moral relativism?

The implications of differential VSLs extend far beyond individual corporate decisions. They shape global trade patterns, environmental regulations, and public health policies (Silverman, 2024; Fioramonti, 2014). When the World Bank evaluates development projects, when international organizations assess environmental initiatives, and when multinational corporations make investment decisions, these VSL disparities influence where pollution-intensive activities end up. A particularly telling example comes from the shipbreaking industry. When aging vessels need to be dismantled, most end up on the beaches of countries with relatively lower VSLs like Bangladesh rather than in specialized facilities in developed nations.

In Bangladesh, one of the world's largest shipbreakers, the expansion of the industry has come at the cost of serious environmental degradation and severe labor exploitation (Alam and Faruque, 2014; Uddin *et al.*, 2024). Researchers have noted how "shipbreaking yards discharged thousands of tons of toxic substances such as asbestos, lead, waste oil and polychlorinated biphenyls (PBCs) into surrounding soils and seawater" (Alam and Faruque, 2014, p. 47), which resulted in many of the former mangrove forests in Bangladesh disappearing completely. Nandi *et al.* (2022) estimated the 2019 VSL of Bangladesh to be about $234,383. In addition, Hossain *et al.* (2016) detailed how asbestos fibers fly freely in the air near the beaches where shipbreaking activities take place, exposure to which can cause serious diseases like cancer and asbestosis. There are also other pollutants involving refrigerants, such as chlorofluorocarbon chemicals, which are hazardous to the ozone layer. These pollutants are dispersed when dismantling the ships' air conditioning and refrigeration systems. In countries like Bangladesh, the cost of proper safety and environmental controls becomes "unnecessary" when using lower VSLs for cost–benefit analyses, making it economically rational to conduct these hazardous operations in developing nations. The Human Rights Watch (2023) quoted Tanvir, a shipbreaker who had been working in the industry since 1982:

> While working in this industry I saw so many of my colleagues lose their lives. But still the system never changed. Workers' rights are violated every day. I think shipbreaking is the most neglected industry in the world.

This silencing of workers' voices — where those who dare speak up about inadequate safety gear face termination, where hands are protected by worn socks against burning metal, where the choice becomes either risking one's life or losing one's livelihood — reflects a system that has institutionalized the expendability of human life.

Perhaps the most troubling aspect of VSL emerges when examining how economic desperation can distort how people value their own lives. While VSL typically measures society's collective willingness to pay for risk reduction, extreme poverty can create situations where individuals make tragic calculations about their lives' monetary worth. This was heartbreakingly illustrated during the 2010 Foxconn crisis in Shenzhen, China (Guo *et al.*, 2012). At these massive electronics manufacturing complexes, young workers faced grueling conditions — 100 hours of monthly overtime, intense isolation despite working shoulder to shoulder, and a prison-like atmosphere of strict regulations and security checkpoints (Blanch, 2010; Merchant, 2017). When Foxconn began offering 100,000 yuan ($14,500) in death compensation to families — nearly ten years' worth of wages for many migrant workers earning just 2,000 yuan monthly — it created what economists term a "perverse incentive."

But the human reality was far more devastating: Some young workers, barely twenty years old, began viewing their deaths as the only way to provide meaningful financial support for their families back home.

The company's solution to widespread despair — a "stress room" where workers could beat dolls with bats (Blanch, 2010) — reflected a profound failure to recognize the human dignity and worth of its workforce. From a cold cost–benefit perspective, Foxconn's actions suggested it valued a worker's life at far less than standard VSL measures used in developed nations. The company appeared to calculate that paying occasional death compensation and installing minimal stress reduction measures were cheaper than fundamentally improving working conditions through reduced hours, higher wages, and better worker support systems — thereby, effectively using a VSL for its workforce that was orders of magnitude below international standards. This crisis revealed how economic systems that devalue human life can lead people to tragically undervalue their own existence, creating a devastating departure from how VSL is meant to reflect society's shared commitment to protecting life. When people live in circumstances where their death might provide more financial security for their families than their continued work, we confront the ultimate failure of our global economic system.

Similar patterns emerge in other contexts. In some pollution-heavy industries, workers accept hazardous conditions without adequate protection because the wage premium, however small, seems worth the health risk. Unlike in developed nations, where wage–risk studies show people demanding substantial compensation for taking on additional mortality risks, workers in developing nations often accept minimal premiums for substantial risks. This "acceptance" of lower self-valuation should not be interpreted as validation of different VSLs. Instead, it highlights how economic desperation can force people to make choices that no one should have to make. When people accept a higher risk of death from air pollution because they need employment, or when workers view their own deaths as a potential economic solution for their families, it reveals the violence of poverty rather than a legitimate difference in how life should be valued.

The devaluation of human life in developing nations isn't limited to current working conditions — it extends into the future through technological transformation. As traditional labor-intensive industries grapple with longstanding inequities, a new wave of technological transformation threatens to deepen these divides even further, creating a double displacement effect — first by hazardous conditions, then by automation. As artificial intelligence and automation reshape industries worldwide, we're witnessing a new form of value stratification. In developed nations, automation often arrives with careful consideration for displaced workers, including retraining programs and social safety nets. However, in developing nations, rapid automation in manufacturing hubs could lead to large-scale displacement without adequate support systems, effectively treating workers as disposable assets rather than human capital worthy of investment (Martins-Neto *et al.*, 2024; Schlogl and Sumner, 2020).

The current practice of linking VSL to GDP per capita raises fundamental philosophical questions: Should a person's life be worth less simply because they were born in a poorer country? If we accept that premise, aren't we implicitly accepting — and perpetuating — global inequality? Proponents of differential VSLs often advance what appears, at first glance, to be a culturally sensitive argument: that using local standards for VSL calculations respects the autonomy and preferences of developing nations. They point to observable differences in risk–wage trade-offs, varied willingness to pay for safety features, and distinct local priorities for resource allocation. This argument suggests that imposing Western valuations of life on other societies represents a form of cultural imperialism, failing to acknowledge legitimate differences in how various societies balance risks and rewards.

Yet this seemingly respectful stance crumbles under closer examination. The fundamental flaw lies in conflating choices made under economic duress with genuine cultural preferences. When a community "accepts" higher pollution levels in exchange for economic opportunity, or when workers "choose" more hazardous conditions for marginally higher pay, these decisions often reflect the constraints of poverty rather than authentic cultural values or preferences. The rhetoric of respecting local standards can serve as a convenient justification for maintaining global inequalities, allowing multinational corporations and policymakers to exploit economic desperation while claiming cultural sensitivity.

Consider how environmental standards evolve as societies become wealthier. Time and again, we observe that once basic economic security is achieved, communities invariably demand stronger environmental protection. This pattern is particularly visible in the evolution of air quality standards. Cities that once accepted smog-filled skies as the price of progress eventually demand clean air as a basic right. Beijing's transformation from accepting heavy industrial pollution to implementing strict air quality controls illustrates how economic growth changes not just the ability to afford environmental protection but the very conception of what constitutes an acceptable risk to human life.

This suggests that conceding to environmental hazards in developing nations likely reflects economic necessity rather than cultural preference. The same communities that "accept" pollution in times of poverty often become its strongest opponents once they achieve greater economic security. This understanding complicates the role of multinational corporations operating in developing regions. When they encounter lower environmental standards or reduced safety expectations, should they adapt to local norms or maintain higher global standards? The easy answer — adapting to local standards — often amounts to exploiting economic vulnerability under the guise of cultural respect. A more ethical approach would recognize that while economic conditions might force people to accept greater risks, this acceptance doesn't morally justify imposing those risks.

Furthermore, the very notion of "local standards" becomes problematic in our interconnected world. When pollution from one region affects global climate patterns, or when industrial practices in one country influence competitive pressures in another, can we really speak of purely local standards? The environmental choices made in developing nations often reflect not just local preferences but global economic pressures and power dynamics.

Perhaps most importantly, the local standards argument fails to account for the vast power differentials in our global economy. When multinational corporations make decisions about where to locate pollution-intensive industries, they're not so much respecting local standards as they are exploiting global inequalities. The "choice" of a community to accept higher pollution levels often occurs within a context of limited alternatives and economic pressure from much more powerful global actors. Moving forward requires acknowledging that true respect for local autonomy means working to expand the real choices available to communities, not simply accepting decisions made under economic duress. It means recognizing that while economic conditions might vary across regions, the fundamental right to breathe clean air and live in a healthy environment should not.

The challenge lies in balancing economic practicality with moral imperatives. While using a single global VSL might seem ethically appealing, it could make many crucial development projects appear economically unfeasible in lower-income countries. Yet the current system of dramatically different VSLs effectively puts a lower price on the lives of the world's most vulnerable people. Perhaps the solution lies not in choosing between these extremes but in rethinking how we value human life entirely. Should economic metrics be the primary basis for such valuations? Could we develop more nuanced approaches that consider both local economic conditions and universal human rights? As the world grapples with global challenges like climate change and air pollution, these questions become increasingly urgent.

The Ford Pinto case sparked outrage because it made explicit what was usually implicit: the practice of putting a price tag on human life. Today's global VSL disparities represent an even larger ethical challenge, one that underpins many decisions about where pollution-intensive industries operate. As we export our environmental burdens to regions with lower VSLs, we must ask ourselves these questions: Are we repeating the Pinto calculation on a global scale? And if so, what does that say about our values as a global society? The answer lies not merely in condemning these cold calculations, but in confronting the bitter truth of how our global economic system has normalized the systematic devaluation of human life based on longitude and latitude. When we trace the invisible lines that connect a comfortable life in one nation to countless shortened lives in another, we must ask ourselves what moral debt accumulates in the ledgers we keep hidden from ourselves. Just as the Pinto case eventually led to stronger automotive safety standards, perhaps understanding today's global VSL disparities can spark a movement toward more equitable international standards for human life and dignity. But this time, we cannot wait for more martyrs to illuminate our conscience.

The challenge before our generation is profound and painful. It demands more than just recognizing these disparities — it requires us to confront our own complicity in a system that prices our breaths differently based on where they're drawn. When a child in Bangladesh breathes in asbestos from a dismantled ship, when workers in Shenzhen views their death as their family's salvation, they fall from heights of dignity that dwarf our own shallow perches. Yet even in this darkness, there are seeds of transformation. Every time we choose to recognize the full humanity in those our economic systems would render invisible, we strike a blow against the machinery of indifference. The road ahead is long and marked by the ruins of many failed reforms and numerous hollow corporate promises, but we must persist. For in this persistence — in refusing to accept that a life's worth can be measured in GDP — lies the only heroism worth pursuing: the courage to face these ugly truths and still believe in our capacity to change them. Our task now is to rebuild not just economic systems but our very understanding of value. We must create frameworks where dignity is not a commodity to be traded but a fundamental right to be protected. When our children look back at this moment, let them see not just our recognition of these disparities but our fierce determination to dismantle them.

References

Alam, S. and Faruque, A. (2014). Legal regulation of the shipbreaking industry in Bangladesh: The international regulatory framework and domestic implementation challenges. *Marine Policy, 47*, 46–56.

Becerra-Pérez, L. A., Ramos-Alvarez, R. A., DelaCruz, J. J., and García-Páez, B. (2024). Value per Statistical Life at the Sub-National Level as a Tool for Assessing Public Health and Environmental Problems. *INQUIRY: The Journal of Health Care Organization, Provision, and Financing, 61*, 469580241246476. DOI: https://doi.org/10.1177/00469580241246476.

Blanch, B. (2010, May 28). Foxconn suicides: 'Workers feel quite lonely'. *BBC News*. Retrieved on February 18, 2024 from https://www.bbc.com/news/10182824.

Dowie, M. (1977, September/October). Pinto Madness. *Mother Jones*. Retrieved on February 18, 2024 from https://www.motherjones.com/politics/1977/09/pinto-madness/.

EPA. (n.d.). Mortality Risk Valuation. *United States Environmental Protection Agency*. Retrieved on February 18, 2024 from https://www.epa.gov/environmental-economics/mortality-risk-valuation.

Fioramonti, D. L. (2014). *How Numbers Rule the World: The Use and Abuse of Statistics in Global Politics*. New York, NY: Zed Books Ltd.

Guo, L., Hsu, S. H., Holton, A., and Jeong, S. H. (2012). A case study of the Foxconn suicides: An international perspective to framing the sweatshop issue. *International Communication Gazette, 74*(5), 484–503.

Hartz, M. (2020, March 12). 1980: Indiana prosecutor charges Ford with reckless homicide following deadly Pinto crash. *WRTV Indianapolis*. Retrieved on February 18, 2024 from https://www.wrtv.com/lifestyle/history/1980-indiana-prosecutor-charges-ford-with-reckless-homicide-following-deadly-pinto-crash.

Hossain, M. S., Fakhruddin, A. N. M., Chowdhury, M. A. Z., and Gan, S. H. (2016). Impact of ship-breaking activities on the coastal environment of Bangladesh and a management system for its sustainability. *Environmental Science & Policy, 60*, 84–94.

Human Rights Watch. (2023, September 28). Trading lives for profit: How the shipping industry circumvents regulations to scrap toxic ships on Bangladesh's beaches. Retrieved on February 18, 2024 from https://www.hrw.org/report/2023/09/28/trading-lives-profit/how-shipping-industry-circumvents-regulations-scrap-toxic.

Martins-Neto, A., Cirera, X., and Coad, A. (2024). Routine-biased technological change and employee outcomes after mass layoffs: Evidence from Brazil. *Industrial and Corporate Change, 33*(3), 555–583.

Merchant, B. (2017, June 18). Life and death in Apple's forbidden city. *The Guardian.* Retrieved on February 18, 2024 from https://www.theguardian.com/technology/2017/jun/18/foxconn-life-death-forbidden-city-longhua-suicide-apple-iphone-brian-merchant-one-device-extract.

Miles, D., Stedman, M., and Heald, A. (2020). Living with COVID-19: Balancing costs against benefits in the face of the virus. *National Institute Economic Review,* 253, R60–R76.

Nandi, A., Counts, N., Chen, S., Seligman, B., Tortorice, D., Vigo, D., and Bloom, D. E. (2022). Global and regional projections of the economic burden of Alzheimer's disease and related dementias from 2019 to 2050: A value of statistical life approach. *EClinicalMedicine, 51,* 101580. DOI: 10.1016/j.eclinm.2022.101580.

Schlogl, L. and Sumner, A. (2020). *Disrupted Development and the Future of Inequality in the Age of Automation.* Cham, Switzerland: Springer Nature.

Silverman, M. (2024). "The value of a statistical life" in economics, law, and policy: Reflections from the pandemic. *Rethinking Marxism, 36*(1), 33–58.

Sweis, N. J. (2022). Revisiting the value of a statistical life: An international approach during COVID-19. *Risk Management, 24*(3), 259–272.

Uddin, M. K., Nobi, M. N., and Islam, A. M. (2024). Environmental hazards and health rights of workers in shipbreaking in Bangladesh. *International Journal of Human Rights in Healthcare, 17*(3), 300–314.

Chapter 4

Shades of Inequality: Air Pollution's Uneven Toll across Communities

On October 23rd, 2024, the US Air Quality Index (AQI) in the Indian capital of New Delhi slipped to 349, the hazardous category (this is labeled as "very poor" in India's labeling system), while some areas even recorded levels above 400 (*HT News Desk*, 2024). However, for resident *Delhiites*, this was nothing new. Just about a year earlier, in the first week of November 2023, Delhi had been blanketed with a toxic haze so thick that it was visible on NASA satellite imagery (Patel, 2023). At that time, the AQI in some places like India Gate had spiked to 999 (Gupta, 2023). Patel (2023) noted that the severe pollution levels forced school closings not just in Delhi but also in neighboring Pakistan. Coaching activities for sports like football and cricket were canceled, and parents had to limit outdoor playtime for children (Mollan, 2023). But staying indoors provided little respite. Yadav and Ghosh (2022) found that indoor air pollution could be quite high in winter months when doors and windows are closed and ventilation is poor. Even when schools reopened mid-November, the AQI remained high around 348, forcing children to bring out the masks that had been put away after the COVID-19 pandemic to venture to class. After a year, it seemed to many *Delhiites* that they were about to face the same old situation.

Although everyone residing in an affected region is exposed to contaminated air, pollution is hardly a great equalizer. Children from low-income families are more adversely impacted by air pollution than those from higher-income households. The homes of the former may not have access to air purifiers or adequate ventilation systems, aggravating the impact on growth, development, and life expectancy (Karkun and Sekar, 2023). Whereas the well-to-do are able to move to cleaner locations, the poor are left with limited options. A stark example of this disparity can be found in Delhi's Bhalswa neighborhood, where thousands live next to a perpetually smoldering landfill (Ellis-Petersen and Hassan, 2024). While affluent areas near Lodhi Garden measure PM2.5 levels around 200, Bhalswa regularly experiences levels exceeding 600, demonstrating how geography and socio-economic status (SES) intersect in environmental exposure. In their interviews, Ellis-Petersen and Hassan (2024) quote resident Mohammad Rizwan on the dumpster in Bhalswa:

> I have watched it grow from a small rubbish heap into that huge mountain over the past 20 years. During the summer it catches fire every week because of all the gas and then it becomes even more disgusting here. It's impossible to breathe and everyone gets sick because of the bad fumes and smoke we have to inhale. It feels so dangerous to live here but I have no choice, this is where my home and livelihood is.

When asked why this grievous waste problem has not been addressed over such a long period of time, Bharati Chaturvedi, founder and director of the Chintan Environmental Research and Action Group, noted the financial factors intricately woven into the considerations (Ellis-Petersen and Hassan, 2024):

> Even the simplest measures for reducing methane are not put in place in Delhi. We need to be composting this waste on a large scale but one big problem is that there's no land to do it. There's also no market for compost, so there's no financial incentive to do anything other than dump organic waste.

In addition, Greenstone *et al.* (2021) found that high SES households are over 13 times more likely to own air purifiers at baseline, compared to households with low SES. Moreover, when schools are closed due to pollution and classes are moved online, children from low SES households may lack access to laptops or an appropriate environment for virtual learning. Even for adults, white-collar workers may choose to work remotely if necessary, but many migrant workers and employees delivering essential services have to brave the toxic haze.

So, how did Delhi get to this point? Since the late 20th century, through a process known as the "race to the bottom," pollution-intensive industries began relocating to countries with relatively lax environmental standards. This phenomenon wasn't merely about seeking cheaper labor; it represented a calculated strategy to exploit regulatory and SES differences between regions. Robert Bullard documented this pattern in his seminal work *Dumping in Dixie* (1990), noting how corporations systematically targeted vulnerable communities for hazardous facilities. This targeting followed clear patterns of path dependency, where historical inequalities created opportunities for further exploitation. China and India's rapid industrialization in the late 20th and early 21st centuries is a prime example. Cities like Beijing and New Delhi became notorious for their hazardous air quality, mirroring the conditions once seen in Western industrial centers. The World Health Organization (WHO) reported in 2016 that 98% of cities in low- and middle-income countries with more than 100,000 inhabitants did not meet WHO air quality guidelines (WHO, 2016). This stark statistic underscores the disproportionate burden borne by developing nations.

The pattern of environmental injustice extends beyond routine industrial pollution to encompass human-made disasters and conflicts. When environmental catastrophes strike — whether through industrial accidents, natural disasters, or human conflict — their impacts are rarely distributed equally. The 1984 Bhopal gas tragedy in India serves as a haunting testament to this reality. When toxic methyl isocyanate gas leaked from the Union Carbide plant, it was the poorest residents living in makeshift settlements near the factory who suffered the most casualties. While middle-class neighborhoods had time to evacuate, the working-class communities, lacking transportation and resources, faced the full brunt of the disaster. Even four decades later, the children and grandchildren of survivors continue to face health complications, while the site remains contaminated (Mukherjee, 2016). Those with fewer resources often face greater exposure and have limited means to protect themselves or relocate to safety. This disparity becomes particularly evident in regions affected by political instability or armed conflicts, where environmental concerns frequently intersect with humanitarian crises.

For example, ongoing wars have created pollution hotspots in regions where previously none existed. When missiles struck Ukraine's industrial facilities in 2022, they released more than just explosive force — they unleashed toxic clouds that recognized no borders or battle lines. The destruction of the Kakhovka Dam in June 2023 not only flooded vast areas but also spread industrial pollutants and chemicals across agricultural lands, creating long-term environmental hazards that will disproportionately affect small farmers and rural communities. Similar incidents near oil refineries, chemical plants, and ammunition depots have created toxic hotspots across Ukraine, disproportionately affecting those too poor or frail to evacuate. Every tank, every fighter jet, every supply convoy added to the carbon burden. A single F-35 fighter jet burns approximately 750 gallons of fuel per hour of flight (Reynolds, 2024), much more than what an average car uses in a whole year. War is inherently carbon intensive. Further, as Europe scrambled to replace Russian gas, countries reverted to coal power — undoing years of progress in emissions reduction. Germany reactivated several coal plants (Connolly, 2022), while Poland increased its coal mining operations. War forced people to choose between energy security and environmental protection, between immediate survival and long-term sustainability.

These global disruptions highlight a fundamental truth about environmental challenges: Their impacts cascade unevenly through society, creating ripple effects that manifest differently across geographic and socio-economic boundaries. The concept of environmental cascades can be observed in the way climate change affects food security. When drought strikes agricultural regions, wealthy nations can simply import food at higher prices, while poorer countries face immediate food shortages. This pattern played out dramatically during the 2017 East Africa drought, where wealthy urban residents could buffer against food price increases, while rural communities faced devastating famine. At that time, a struggling survivor noted the following (Oxfam International, 2017):

> We have moved four times in the last four months. We were trying to follow the rain — moving according to where the rains were supposed to come. But they haven't. If the rains don't come, none of us will survive.

While war represents an extreme example of human-induced environmental degradation, similar patterns of disproportionate impact can be observed in peacetime contexts, where economic and social inequalities shape exposure to environmental hazards.

Within countries, the impact of air pollution is far from uniform. In both developed and developing nations, poorer communities often bear the brunt of environmental degradation. In the United States, for instance, studies have shown that low-income neighborhoods and communities of color are more likely to be located near polluting industries or high-traffic areas. The case of "Cancer Alley" in Louisiana, where predominantly African American communities live amid one of the country's highest concentrations of petrochemical plants, exemplifies this pattern. Residents there face cancer risks multiple times higher than the national average because of the emission of the highly potent toxin ethylene oxide (Sadasivam, 2024), yet economic dependence on these industries creates a cruel choice between livelihood and health.

For those living in these sacrifice zones, the weight of environmental injustice manifests not just in physical ailments but in a profound spiritual exhaustion. Factory workers in Cancer Alley often feel an overwhelming sense of fatalism — a crushing recognition that their lives are somehow deemed less valuable than others. They wake up every morning knowing the air they breathe is a slow killer, yet they still have to go to work at the same plants that are poisoning their bodies. This cruel paradox — being forced to participate in one's own destruction for survival — creates a unique form of psychological trauma. Workers feel trapped in an endless cycle, where each day's labor contributes to the toxic burden on their health. This existential weight transforms everyday decisions into pro-found moral dilemmas: Is it better to work overtime at the plant to afford an air purifier for your children's room, knowing those extra hours of exposure might mean fewer years to spend with them? Such questions have no clear answers, leaving communities to navigate an impossible maze of choices between immediate survival and long-term existence. In these communities, we can see a particular kind of mourning — not for what has been lost, but for what never had the chance to be.

This pattern of sacrificing vulnerable communities repeats itself across contexts, whether through air pollution from chemical plants or, as in Flint, Michigan, through contaminated water systems. In both cases, the fundamental dynamic remains the same — communities with the least political power are forced to bear the greatest environmental burdens, their bodies becoming unwilling ledgers of corporate and governmental negligence. In developing countries, the disparity is even more pro-nounced. In India, lower-income communities in Delhi were exposed to significantly higher levels of air pollution compared to wealthier areas (Chatterjee, 2019). Similarly, in China, rural-to-urban migrants often reside in areas with poorer air quality and have relatively limited access to medical care to address resulting health issues. These migrants form the backbone of China's urban workforce yet are systematically excluded from the environmental protections afforded to wealthier residents.

The problem of air pollution is not just about the adverse impact on health but also about the loss of the full human experience of being alive in the world. This loss of sensory experience has profound psychological implications — studies have shown increased rates of depression, anxiety, and suicides in heavily polluted areas, creating a feedback loop of mental and physical health challenges (Gladka *et al.*, 2018; Ventriglio *et al.*, 2021; and Yang *et al.*, 2023). In cities like Delhi, many will never know the taste of rain that hasn't passed through layers of pollution, never experience the sharp clarity of winter air untainted by exhaust fumes, and never smell the brisk clean wind carrying the scent of ripening fruits. During Diwali, for instance, the ancient practice of lighting diyas (oil lamps) is now often obscured by heavy smog, disconnecting people from centuries-old traditions. In Nepal's Kathmandu Valley, once celebrated for its crisp mountain air and where tourists come seeking the pure Himalayan experience, air pollution has significantly reduced the visibility of the mountains compared to the 1970s (Mandal, 2021). Local artists who traditionally painted mountain landscapes now struggle to even see their subjects. In addition, when the air becomes dangerous, people become fearful and retreat from public spaces. The spontaneous human interactions that flavor daily life disappear. Children who might have played cricket in the streets until dusk now stay indoors, their social development curtailed by invisible dangers. As we lose the ability to taste the essential flavors of existence, we lose part of what makes us human.

Like Truman Burbank in Andrew Niccol's *The Truman Show*, who gradually realizes his world is an orchestrated fiction, communities bearing the heaviest burden of air pollution are awakening to the carefully constructed narratives about economic progress and shared sacrifice. This awakening often follows a similar pattern to Truman's — small inconsistencies pile up until they can no longer be ignored. In the stretch of the Mississippi River between New Orleans and Baton Rouge in Louisiana, residents began noticing unusually high cancer rates long before official studies confirmed their fears. Just as Truman's world was controlled by unseen directors, vulnerable populations find their environmental reality shaped by distant decision-makers who themselves breathe cleaner air. The parallel extends even to the way information is controlled — just as Truman's media was carefully curated, communities often face deliberate misinformation about pollution levels and health risks. The residents of Love Canal, New York, for instance, were repeatedly told their neighborhood was safe despite sitting atop thousands of tons of toxic waste. Similar to how Truman faces artificial storms when he tries to escape, communities attempting to challenge environmental inequality often encounter manufactured obstacles. In Warren County, North Carolina, when residents protested against toxic waste dumping in their predominantly African American community in 1982, they faced not just police resistance but also carefully crafted narratives about economic necessity and scientific safety — their own version of Truman's artificial reality.

This environmental staging plays out globally. In Delhi's poorest neighborhoods, where PM2.5 levels can reach 10 times those in diplomatic enclaves, residents are often told this is the price of development, and that everyone is in this together. This narrative ignores the fact that while air pollution affects everyone, its intensity varies dramatically by location and SES. Just as Truman's show creators insisted they were providing him with a better life, industry representatives and politicians often argue that pollution is a necessary price for economic progress. Although India had a nominal GDP of over $3.8 trillion in 2024, Greenstone and Hanna (2014) found that India's particulate matter exposure was more than five times that of the United States and that air pollution was responsible for more than 12.5% of all deaths in India. Like the fake rain machine in Truman's dome, authorities often deployed theatrical solutions that created an illusion of action while the true crisis continued. Delhi's $2 million smog tower built in 2021 stands as a perfect metaphor for this environmental theater — a spectacular prop in the ongoing show of environmental concern. With its 40 giant fans reaching toward the blurred sky, the tower represents performative environmentalism — actions designed to be seen rather than to solve. The tower itself runs on electricity generated by the very coal plants choking Delhi.

Even more misleading is the government's rephrasing of air quality classifications, creating its own version of Truman's weather report. While the U.S. Air Quality Index labels an AQI over 300 as "hazardous," Delhi's Graded Response Action Plan (GRAP) for the National Capital Region (NCR) and adjoining areas euphemistically terms it "very poor" or "severe" (CAQM, 2024), normalizing the unacceptable and thereby numbing residents to danger. This linguistic manipulation has real consequences — when schools remain open on days labeled "very poor" instead of "hazardous," children's health is compromised by semantic games. India's standards have been adapted to accommodate routinely poor air quality rather than adhering to international health-based benchmarks. When "severe" becomes normal, when deadly air is just another Tuesday, people stop fighting back. Like Truman accepting the artificial sun, people become conditioned to accept toxicity as natural.

This carefully constructed narrative of acceptable pollution levels, combined with highly visible but ineffective solutions, creates a spectacle of action — a performance that actually prevents real change by satisfying the public's desire to see something, anything, being done. The implementation of odd–even vehicle rationing schemes in Delhi, while visually impressive and widely publicized, has had a minimal impact on overall air quality. Yet it serves its theatrical purpose — creating the impression of decisive action while avoiding more fundamental but politically difficult solutions like addressing industrial emissions or agricultural burning. Residents in these highly polluted areas often develop a form of learned helplessness similar to Truman's initial acceptance of his artificial world. However, like Truman's gradual awakening, communities need to recognize and challenge the *status quo*. When people realize their reality is being controlled, that their children's lives are being sacrificed for someone else's profit, they cannot unsee it. Like Truman hitting the wall of his world to escape, communities have to rise up to hit the wall of their lies.

In *How the Steel Was Tempered*, author Nikolai Ostrovsky (1934/1981) described the journey of the protagonist Pavel Korchagin, who gave his health to build socialism's foundation. Unlike Pavel's conscious choice, residents of pollution-heavy zones make involuntary sacrifices for others' prosperity. The novel's central metaphor — the tempering of steel requiring intense heat and pressure — finds a grim parallel in how poverty and political powerlessness "temper" communities into accepting environmental devastation. Just as steel must endure extreme conditions to gain strength, these communities are forced to endure extreme pollution to survive economically. The difference lies in agency — while Korchagin chose his sacrifice for a believed greater good, modern communities with low SES are often sacrifice zones by circumstance rather than choice. As we continue to grapple with global environmental inequalities, Korchagin's vision of conscious sacrifice for collective good stands in stark contrast to our current reality of imposed sacrifice for private gain. The question becomes not how the steel was tempered, but who benefits from the tempering, and at what cost to human dignity and life itself.

References

CAQM. (2024). GRAP schedule. *Commission for Air Quality Management in National Capital Region and Adjoining Areas.* Retrieved on October 20, 2024 from https://caqm.nic.in/index1.aspx?lsid=4217&lev=2&lid=4216&langid=1.

Chatterjee, P. (2019). Indian air pollution: Loaded dice. *The Lancet Planetary Health, 3*(12), e500–e501.

Connolly, K. (2022, July 8). Germany to reactivate coal power plants as Russia curbs gas flow. *The Guardian.* Retrieved on October 20, 2024 from https://www.theguardian.com/world/2022/jul/08/germany-reactivate-coal-power-plants-russia-curbs-gas-flow.

Ellis-Petersen, H. and Hassan, A. (2024, February 12). 'It's impossible to breathe': Delhi's rubbish dumps drive sky-high methane emissions. *The Guardian.* Retrieved on October 20, 2024 from https://www.theguardian.com.

Gładka, A., Rymaszewska, J., and Zatoński, T. (2018). Impact of air pollution on depression and suicide. *International Journal of Occupational Medicine and Environmental Health, 31*(6), 711–721.

Greenstone, M. and Hanna, R. (2014). Environmental regulations, air and water pollution, and infant mortality in India. *American Economic Review, 104*(10), 3038–3072.

Greenstone, M., Lee, K., and Sahai, H. (2021). Indoor air quality, information, and socioeconomic status: Evidence from Delhi. In *AEA Papers and Proceedings* (Vol. 111), Nashville, TN 37203: American Economic Association, pp. 420–424.

Gupta, A. (2023, November 13). Diwali hangover: Delhi AQI crosses hazardous levels of 1000 in many areas as firework ban goes up in flames. *The Weather Channel.* Retrieved on April 15, 2024 from https://weather.com/en-IN/india/pollution/news/2023-11-13-delhi-aqi-touches-999-after-diwali-despite-firework-ban.

HT News Desk. (2024, October 23). Delhi AQI recorded at 349, slips into 'very poor' category amid rising air pollution; schools to be closed soon? *Hindustan Times.* Retrieved on October 24, 2024 from https://www.hindustantimes.com/india-news/delhi-air-pollution-aqi-today-october-23-2024-349-slips-into-very-poor-category-101729649691324.html.

Karkun, A. and Sekar, A. (2023, November 28). Why closing schools does not protect children from air pollution. *Scroll.* Retrieved on April 15, 2024 from https://scroll.in/article/1059479/why-closing-schools-does-not-protect-children-from-air-pollution.

Mandal, K. (2021, March 27). Haze and smoke turn Kathmandu most polluted, low visibility affects flights. *Kathmandu Post.*

Mollan, C. (2023, November 6). Delhi pollution: No school, no play for city's children. *BBC News.* Retrieved on April 15, 2024 from https://www.bbc.com/news/world-asia-india-67330016.

Mukherjee, R. (2016). Toxic lunch in Bhopal and chemical publics. *Science, Technology, & Human Values, 41*(5), 849–875.

Ostrovsky, N. (1934/1981). *How the Steel was Tempered.* Moscow, Russia: Progress Publishers.

Oxfam International. (2017, April 27). A climate in crisis: How climate change is making drought and humanitarian disaster worse in East Africa. *Oxfam Media Briefing.* Retrieved on April 15, 2024 from https://www.oxfam.org/en/research/climate-crisis.

Patel, K. (2023, November 9). The smog choking this Indian city is visible from space. *The Washington Post.* Retrieved on April 15, 2024 from https://www.washingtonpost.com/climate-environment/2023/11/09/india-air-quality-smog-new-delhi/.

Reynolds, A. (2024, October 2). Understanding Fuel Consumption in Modern Military Aviation: A Look at the F-35. *BlackBird Media S.A.* Retrieved on October 20, 2024 from https://agogs.sk.

Sadasivam, N. (2024, June 12). Real-time data show the air in Louisiana's 'Cancer Alley' is even worse than expected. *Louisiana Illuminator.* Retrieved on October 20, 2024 from https://lailluminator.com/2024/06/12/cancer-alley-4/.

Ventriglio, A., Bellomo, A., di Gioia, I., Di Sabatino, D., Favale, D., De Berardis, D., and Cianconi, P. (2021). Environmental pollution and mental health: A narrative review of literature. *CNS spectrums, 26*(1), 51–61.

Yadav, A. K. and Ghosh, C. (2022). Spatiotemporal variation of particulate matter & risk of exposure in the indoor-outdoor residential environment: a case study from urban city Delhi, India. *Pollution, 8*(3), 860–874.

Yang, T., Wang, J., Huang, J., Kelly, F. J., and Li, G. (2023). Long-term exposure to multiple ambient air pollutants and association with incident depression and anxiety. *JAMA psychiatry, 80*(4), 305–313.

WHO. (2016). Air quality database: Update 2016. *World Health Organization.* Retrieved on April 14, 2024 from https://www.who.int/data/gho/data/themes/air-pollution/who-air-quality-database/2016.

Chapter 5

Digital Haze Ahead: The Hidden Environmental Cost of Cloud Computing

In the summer of 2024, some residents of Maiden, North Carolina, woke to a familiar low ambient sound — the constant whir of cooling fans from Apple's nearby data center. The 500,000-square-foot facility, which consumed millions of gallons of water daily to keep its servers cool, had become an emblem of a growing crisis. Even as local farmers faced water use restrictions due to drought conditions across the state (North Carolina Environmental Quality, 2024), the data center's cooling systems continued running — each swipe, scroll, and cloud backup from millions of iPhones demanding its share of the region's dwindling water supply. The situation in Maiden wasn't unique; similar scenarios were playing out across the country and around the world. These water-intensive operations were particularly troubling in regions already grappling with climate change-induced water scarcity.

The irony wasn't lost on Maiden's residents. The same facility that promised to herald their small town's entry into the digital age was now competing with them for basic resources. The data center, despite its pristine exterior and promises of renewable energy (Gooding, 2024), was consuming the equivalent of a small city's water supply (Sverdlik, 2016) — all to keep the virtual world running while the physical one struggled with drought. This stark contrast between digital promises and environmental reality echoed in similar situations worldwide. In Singapore, where data centers had intensively utilized limited land, water, and energy resources, alarmed authorities implemented a moratorium on new data center construction in 2019, citing unsustainable resource consumption patterns (Ng *et al.*, 2023). Subsequently, the moratorium was lifted in 2022, and in May 2024, the country unveiled a plan for additional capacity for "green" data centers, endeavoring to provide a roadmap for sustainable growth of the industry (Mah, 2024). The Singapore case exemplifies the growing tension between digital infrastructure expansion and environmental sustainability. If the previous chapter revealed how the burden of visible pollution falls unequally on vulnerable communities, our digital age presents an even more insidious form of environmental inequality — one cloaked in the promise of clean technology and virtual progress. Like the carefully constructed narrative in *The Truman Show*, the tech industry has created its own compelling fiction — that our transition to a digital world means freedom from industrial pollution's grimy reality.

This digital sleight of hand rivals the environmental theater of Delhi's smog towers. Just as authorities deploy spectacular but ineffective solutions for visible air pollution, the tech industry promotes cloud computing and digitalization as an environmental panacea. The reality, however, tells a different story. Our seemingly weightless digital world has very physical foundations that perpetuate and sometimes worsen the environmental inequalities discussed previously. We can say that similar to greenwashing, we are seeing something like cloud-washing, where firms present cloud and digital services as inherently environmentally friendly while obscuring their physical impact. In this regard, O'Brien (2024) noted that an in-depth analysis over 2020–2022 revealed "the real emissions from the 'in-house' or company-owned data centers of Google, Microsoft, Meta and Apple are probably about 662% — or 7.62 times — higher than officially reported." O'Brien (2024) described how tech companies engage in creative accounting, instead of location-based accounting, to offset in-house emissions. They often do this by purchasing Renewable Energy Certificates (RECs) to calculate market-based emissions, which may be vastly different from actual emissions. Bergamo (2023) succinctly detailed how RECs are being misused by firms to claim that most or all of their energy comes from renewable resources:

> The most puzzling aspect of RECs is the fact that they do not represent the actual electricity that results from green energy production — only its social and environmental components... This company, however, is legally allowed to make claims such as "100% of energy comes from renewable sources" — even if the actual electricity was produced by a coal plant.

Departing from their original intention, RECs have now increasingly become a tool for corporate messaging, instead of facilitating concrete gains in decarbonization.

The creative accounting and obscuration of real environmental impacts become particularly concerning when we consider the vast infrastructure required to power our virtual existence. Data centers around the world, the factories of our digital age, now account for up to 4% of global energy consumption, with the numbers growing at an unprecedented rate (Patnaik, 2024). This rise is fueled by the increase in demand for technologies related to artificial intelligence. A report by Goldman Sachs (2024) stated that "on average, a ChatGPT query needs nearly 10 times as much electricity to process as a Google search." Some estimates suggest that training a single large language model can produce as much carbon dioxide as five cars over their entire lifetimes (Hao, 2019). On that note, the Electric Power Research Institute in the United States estimated that data centers could use up to 9% of the total electricity generated in the country by the end of the decade (Kearney, 2024). At the same time, the International Energy Agency observed that worldwide, data center electricity usage is set to double by 2026 (IEA, 2024). In this regard, Goldman Sachs (2024) predicted that the carbon dioxide emissions of data centers could also more than double between 2022 and 2030. The exponential growth in data center energy consumption is particularly concerning when considering the emergence of quantum computing and advanced AI systems, which could potentially increase energy demands by orders of magnitude.

The world's digital factories often cluster in regions with cheap electricity and lax environmental standards. The parallel to Warren County's historical toxic waste struggle is striking — communities near massive data center complexes face a new kind of environmental burden, one masked by the pristine facades of tech campuses but manifested in the constant drone of cooling systems and the indirect pollution from power generation. In places like Ashburn, Virginia, dubbed the "Data Center Alley" or the "Dulles Technology Corridor" (Woolridge, 2021), local communities experience a modern version of the environmental staging discussed in the previous chapter. Just as Delhi's residents are told that pollution is the price of progress, communities near data centers are assured that the constant noise, increased energy demand, and indirect environmental impact are necessary sacrifices for digital progress. The concentration of data centers in specific regions has led to significant local challenges, including decreased property values in adjacent neighborhoods, groundwater depletion, and strain on the local power grid (Osaka, 2023; Taha and Olufemi, 2023).

The environmental impact extends beyond just operational concerns. The manufacturing and disposal of data center equipment create additional environmental burdens, often borne by communities far from where the digital services are consumed. Major technology companies like Alphabet, Amazon, Microsoft, and Meta frequently invest in and upgrade to state-of-the-art hardware to remain competitive, resulting in fast replacement cycles and the generation of significant volumes of e-waste. The production of servers and networking equipment for data centers also generates significant electronic waste, with much of it ending up in developing countries, despite international agreements like the Basel Convention attempting to regulate such transboundary transfers (Ackah, 2017). In e-waste recycling sites like Agbogbloshie, Ghana, workers had to endure harsh conditions and very high levels of PM2.5-10 concentrations caused by the transportation and burning of these materials, aggravating the accumulation of toxic compounds in their bodies (Awere *et al.*, 2020). In this regard, Maphosa (2022) called on developing countries to "craft policies that reduce the effects of e-waste on the environment by managing data center e-waste and increasing awareness" (p. 21). This highlights the importance of tracking and actively managing the carbon footprint of the entire life-cycle of digital infrastructure.

The cryptocurrency boom provides perhaps the starkest example of this digital environmental theater, accentuating how the pursuit of digital assets can lead to very physical environmental consequences. In Inner Mongolia, where coal remains the primary power source, Bitcoin mining operations have created new pollution hotspots (Xie, 2022), reminiscent of the industrial migrations discussed previously. These Bitcoin mines are essentially large data centers with thousands of computers. Cheng (2024) pointed out that one coal-powered Bitcoin mining site is responsible for 8000–13,000 kg of carbon dioxide emissions per Bitcoin it mines, and that between 2017 and 2021, such mining activities contributed to approximately 77.84 million tons of carbon dioxide emissions in China. In China, which claims about 70% of the global hash rate for mining Bitcoin — that is, the total combined computational power for verifying transactions on the blockchain's ledger — mining operations tend to concentrate in a few places with lower rent, better climate, and cheap electricity like Inner Mongolia, Xinjiang, and Sichuan (Barrett, 2021). After the Chinese government's repeated crackdown on the industry, many Bitcoin mining operations moved to Kazakhstan — a country with a relatively loose regulatory environment, making it the second-largest producer of Bitcoin in the world in 2021 (Pandey, 2023). When mining intensified in the city of Ekibastuz and nearby regions, the facilities started consuming unsustainable amounts of energy, several times the peak demand of the city itself (Guest, 2023). Even in countries with cleaner energy grids, cryptocurrency mining has strained local power systems and led to concerns about the displacement of other industries that could use renewable energy more efficiently.

Yet, unlike the visible smog that blankets Delhi, or the toxic clouds released by warfare in Ukraine, the environmental impact of our digital lives remains largely invisible to end users. When we stream videos, send emails, or store photos in the cloud, we don't see the coal being burned to power these services. This invisibility creates a new form of psychological distancing — one that allows us to maintain the illusion of clean technology while its environmental costs accumulate in communities far from our awareness. It leads to a new form of digital carbon footprint blindness, wherein the abstract nature of digital services makes it difficult for users to connect their regular online activities, like making queries on ChatGPT, with real-world environmental impacts. This disconnect is particularly pronounced in emerging technologies like virtual reality and the metaverse, where the immersive nature of the experience further obscures its physical resource demands. The development of metaverse infrastructure alone, including data centers, graphics processing units, and networking, could consume as much energy as a small European country (Stoll *et al.*, 2022).

However, the tech industry's response to growing environmental scrutiny suggests a possible departure from the "race-to-the-bottom" pattern described earlier. Unlike traditional industries that simply relocated to avoid environmental regulations, some tech companies are investing in genuine solutions. Microsoft's Project Natick, which explored underwater data centers cooled by ocean currents, demonstrated significantly better power usage effectiveness than traditional facilities while reducing water consumption. Google's DeepMind AI system, which optimizes data center cooling, has achieved significant reductions in cooling energy costs (Evans and Gao, 2016). Amazon's commitment to 100% renewable energy, while still a work in progress, shows how digital infrastructure could potentially break free from the cycles of environmental inequality. These innovations represent genuine attempts to address environmental impacts rather than merely relocating or obscuring them.

Beyond innovative cooling solutions, the tech industry has begun exploring more radical approaches to reducing data center environmental impact. The concept of circular data centers has gained traction, with Microsoft opening its pilot Circular Center in Amsterdam in 2020, aimed at processing decommissioned cloud servers and hardware, sorting, and intelligently channeling the components and equipment to optimize reuse or repurpose to move "toward its goal of reusing 90% of its cloud computing hardware assets by 2025" (Borkar, 2022). In a detailed report, Kairos Future (2024) highlighted the urgent need to invest in circular solutions because of the massive growth in e-waste (p. 20):

> A trip back to the early 2000s would likely remind us of all the objects in our lives that used to not be electronics — much less smart electronics. Watches, bikes, furniture, glasses and even cars. With booming semiconductor and microchips industries, there have been incentives for producers to integrate smart functions into an ever-growing number of objects.

To address the situation and accelerate innovation to support a circular electronics industry, strategic alliances have been formed, like the *Circular Electronics Partnership*, which offers a global collaborative platform for interested businesses, and the *Circular Electronics Initiative*, which is an international network with more than 30 member organizations seeking to encourage stakeholders to use and manage electronics in a more circular way.

Yet significant challenges remain. The proliferation of AI-powered applications and smart devices in everyday life, from virtual assistants to autonomous vehicles, has created a paradox where efficiency gains in individual systems are offset by the exponential growth in AI deployment. The explosive growth of data-hungry innovations threatens to overwhelm even the most ambitious efficiency improvements. This is particularly concerning because data centers typically rely on backup power systems, often diesel generators, which activate during peak demand or grid instability. In regions with unreliable power infrastructure, these generators can run for hundreds of hours annually, releasing significant amounts of nitrogen oxides and particulate matter into nearby communities. Furthermore, the increased energy demand from data centers often forces utility companies to rely more heavily on fossil fuel power plants during peak usage periods, contributing to regional air quality degradation. Moreover, the benefits and burdens of digital infrastructure continue to be distributed unequally, with many communities lacking the resources to protect themselves from the indirect environmental impacts of our digital transformation. The emergence of edge computing — which facilitates instant access to data — while promising reduced latency and improved service delivery, risks creating new environmental hotspots as data processing moves closer to end users. This could potentially exacerbate existing environmental justice issues, particularly in urban areas already struggling with heat island effects and air quality concerns.

Looking ahead, the integration of environmental concerns into digital infrastructure planning could catalyze broader changes in how we approach technological development. The emergence of "green coding" practices (Vartziotis *et al.*, 2024), which prioritize energy-efficient software design, and the growing movement toward digital sobriety — the conscious limitation of unnecessary data processing and storage — suggest possible paths toward more sustainable digital futures. However, realizing these possibilities will require sustained attention to transparency, community engagement, and genuine innovation — rather than mere environmental theater — alongside a continuous commitment to ensuring that the benefits of digital innovation do not come at the expense of vulnerable communities.

References

Ackah, M. (2017). Informal E-waste recycling in developing countries: Review of metal(loid)s pollution, environmental impacts and transport pathways. *Environmental Science and Pollution Research, 24*(31), 24092–24101.

Awere, E., Obeng, P. A., Bonoli, A., and Obeng, P. A. (2020). E-waste recycling and public exposure to organic compounds in developing countries: A review of recycling practices and toxicity levels in Ghana. *Environmental Technology Reviews, 9*(1), 1–19.

Barrett, E. (2021, May 1). A Chinese province powered 8% of all Bitcoin mining. Then the government gave miners 2 months to get out. *Fortune.* Retrieved on September 13, 2024 from https://fortune.com/2021/05/02/bitcoin-mining-hashrate-china-inner-mongolia-ban/.

Bergamo, E. (2023, June 15). Renewable Energy Credits: Decarbonizing the grid or just a corporate messaging tool? *Kleinman Center for Energy Policy.* Retrieved on September 13, 2024 from https://kleinmanenergy.upenn.edu/commentary/blog/renewable-energy-credits-decarbonizing-the-grid-or-just-a-corporate-messaging-tool/.

Borkar, R. (2022, March 15). Learn how Microsoft Circular Centers are scaling cloud supply chain sustainability. *Microsoft.* Retrieved on September 13, 2024 from https://azure.microsoft.com/en-us/blog/learn-how-microsoft-circular-centers-are-scaling-cloud-supply-chain-sustainability/.

Cheng, K. Y. (2024). China's crackdown on crypto mining from a climate perspective: Unified efforts from administrative authorities and the judiciary. *International Journal of Digital Law and Governance, 1*(1), 91–112.

Evans, R. and Gao, J. (2016, July 20). DeepMind AI Reduces Google Data Centre Cooling Bill by 40%. *Google DeepMind.* Retrieved on September 13, 2024 from https://deepmind.google/discover/blog/deepmind-ai-reduces-google-data-centre-cooling-bill-by-40/.

Goldman Sachs. (2024, March 14). AI is poised to drive 160% increase in data center power demand. Retrieved on September 13, 2024 from https://www.goldmansachs.com/insights/articles/AI-poised-to-drive-160-increase-in-power-demand.

Gooding, M. (2024, April 19). Apple deploys novel data center air filter that cuts waste and energy use. *Data Centre Dynamics.* Retrieved on September 13, 2024 from https://www.datacenterdynamics.com/en/news/apple-deploys-novel-data-center-air-filter-that-cuts-waste-and-energy-use/.

Guest, P. (2023, January 12). Bitcoin mining was booming in Kazakhstan. Then it was gone. *MIT Teaching Review.* Retrieved on September 13, 2024 from https://www.technologyreview.com/2023/01/12/1066589/bitcoin-mining-boom-kazakhstan/.

Hao, K. (2019). Training a single AI model can emit as much carbon as five cars in their lifetimes. *MIT Technology Review.* Retrieved on September 13, 2024 from https://www.technologyreview.com/2019/06/06/239031/training-a-single-ai-model-can-emit-as-much-carbon-as-five-cars-in-their-lifetimes/.

IEA. (2024). Electricity 2024: Analysis and forecast to 2026. *International Energy Agency.* Retrieved on September 13, 2024 from https://www.iea.org/reports/electricity-2024.

Kairos Future. (2024). The landscape of circular electronics: Towards 2035. *Commissioned by Circular Electronics Initiative.* Retrieved on September 13, 2024 from https://tcocertified.com/circular-electronics-day-case/the-future-of-circular-electronics-new-trend-report-reveals-key-insights-towards-2035/.

Kearney, L. (2024, May 29). Data centers could use 9% of US electricity by 2030, research institute says. *Reuters.* Retrieved on September 13, 2024 from https://www.reuters.com/business/energy/data-centers-could-use-9-us-electricity-by-2030-research-institute-says-2024-05-29/.

Mah, P. (2024, June 25). Singapore lays the groundwork for smart data center growth. *Data Center Dynamics.* Retrieved on September 13, 2024 from https://www.datacenterdynamics.com/en/analysis/singapore-lays-the-groundwork-for-smart-data-center-growth/.

Maphosa, V. (2022). Sustainable e-waste management at higher education institutions' data centers in Zimbabwe. *International Journal of Information Engineering and Electronic Business, 14*(5), 15–23.

Ng, A., Choo, Y.M., Cope, R., and Hilton, A. (2023, August 30). Lessons from Singapore data centers. *Ashurst.* Retrieved on September 13, 2024 from https://www.ashurst.com/en/insights/lessons-from-singapore-data-centres/.

North Carolina Environmental Quality. (2024, June 27). Drought, dry conditions impacting 99 North Carolina counties. *NCEQ Press Releases.* Retrieved on September 13, 2024 from https://www.deq.nc.gov/news/press-releases/2024/06/27/drought-dry-conditions-impacting-99-north-carolina-counties.

O'Brien, I. (2024, September 15). Data center emissions probably 662% higher than big tech claims. Can it keep up the ruse? *The Guardian.* Retrieved on September 17, 2024 from https://www.theguardian.com/technology/2024/sep/15/data-center-gas-emissions-tech.

Osaka, S. (2023, April 25). A new front in the water wars: Your internet use. *The Washington Post.* Retrieved on September 17, 2024 from https://www.washingtonpost.com.

Pandey, (2023, May 5). The rise and fall of the Kazakhstan bitcoin mining industry. *Be(in)Crypto.* Retrieved on September 13, 2024 from https://beincrypto.com/.

Patnaik, C. (2024, March 22). Data center power: Fueling the digital revolution. *Data Center Knowledge.* Retrieved on September 13, 2024 from https://www.datacenterknowledge.com/energy-power-supply/data-center-power-fueling-the-digital-revolution.

Stoll, C., Gallersdörfer, U., and Klaaßen, L. (2022). Climate impacts of the metaverse. *Joule, 6*(12), 2668–2673.

Sverdlik, Y. (2016, June 15). How much water do apple data centers use? *DataCenter Knowledge.* Retrieved on September 13, 2024 from https://www.datacenterknowledge.com/hyperscalers/how-much-water-do-apple-data-centers-use-.

Taha, A. and Olufemi, A. (2023, November 16). Data centers 'straining water resources' as AI swells. *Science X Network.* Retrieved on September 13, 2024 from https://phys.org/news/2023-11-centers-straining-resources-ai.html.

Vartziotis, T., Dellatolas, I., Dasoulas, G., Schmidt, M., Schneider, F., Hoffmann, T., Kotsopoulos, S., and Keckeisen, M. (2024, April). Learn to code sustainably: An empirical study on green code generation. *Proceedings of the 1st International Workshop on Large Language Models for Code,* pp. 30–37.

Woolridge, G. (2021, November 12). How Ashburn, VA became the Colocation Mecca known as Data Center Alley. *Lightyear*. Retrieved on September 13, 2024 from https://lightyear.ai/blogs/ashburn-colocation-data-center-alley.

Xie, B. (2022). Environmental consequences of mining bitcoin: The carbon emission in China. *Highlights in Science, Engineering and Technology, 26,* 41–45.

A Tale of Two Countries:
Blue Skies Lost and Recovered

Chapter 6

China's Blue Sky Recovery: A Journey from Smog to Sustainability

On April 16, 2006, Beijing vanished beneath an immense yellow shroud. In what meteorologists would call the worst sandstorm in years, howling winds swept a staggering 330,000 tons of sand across the capital in a single night (*China Daily*, 2006). The storm was so vast it covered one-eighth of China's territory. As visibility plummeted and the air turned a murky yellow-brown, Beijing's pollution levels hit Grade V — the highest on the scale. This was the eighth sandstorm to hit Beijing that year alone, part of a worsening environmental crisis driven by prolonged droughts, rising temperatures, and rapid desertification in Inner Mongolia. The city, which had promised 230 "blue-sky" days for its upcoming 2008 Olympics, had seen only 56 such days that year — 16 fewer than the same period in the previous year. As communities resorted to washing rather than sweeping their dust-covered roads, the China Meteorological Administration warned that three more sandstorms were yet to come before the month's end.

The late 20th century marked a period of unprecedented economic growth for China, driven by industrialization and urbanization. This growth was fueled, in part, by the offshoring of pollution-intensive industries from Western countries to China. The global rise of mass consumption further increased production pressures on Chinese industries, exacerbating air pollution in many industrial cities. By the 1990s and early 2000s, air pollution in China had reached critical levels, with cities like Beijing, Shanghai, and Guangzhou frequently experiencing hazardous air quality. Particulate matter (PM2.5) levels far exceeded World Health Organization (WHO) guidelines, posing severe health risks to residents.

The Chinese government's initial response to this crisis focused primarily on end-of-pipe solutions and the relocation of heavy industries away from urban centers. However, these measures proved insufficient in addressing the scale and complexity of the problem. A significant turning point in China's approach to air pollution control came with the preparations for the 2008 Beijing Olympics. This event catalyzed a series of stringent measures aimed at rapidly improving air quality in the capital city.

The Beijing Olympics case provides valuable insights into the potential effectiveness of concerted, multi-pronged efforts to combat air pollution. In the lead-up to the Games, authorities implemented a comprehensive strategy that included the temporary closure of factories in and around Beijing, stringent restrictions on vehicle use through an odd–even license plate system, halting of construction activities, and accelerated adoption of stricter vehicle emission standards (Liu and Ogunc, 2023). These measures resulted in a marked improvement in Beijing's air quality during the Olympics, with PM2.5 levels decreasing significantly compared to the previous year. While temporary, this improvement demonstrated the potential for effective pollution control strategies when implemented with political will and adequate resources.

The transformative impact of the 2008 Olympics continued to influence China's environmental policies, particularly in preparation for the 2022 Winter Olympics. While the 2008 Games demonstrated the potential for short-term improvements, the lead-up to the 2022 Games reflected China's more comprehensive, long-term approach to air quality management. After pollution reached crisis levels in 2013, becoming a source of international scrutiny and domestic discontent, Chinese authorities pledged to combat pollution "with an iron fist" (Associated Press, 2022). This commitment coincided with Beijing's bid for the Winter Games and led to more systematic changes. The measures echoed those of 2008 but were implemented on a broader scale, including stricter emission standards for coal-fired plants, vehicle restrictions to reduce emissions, and the replacement of coal-fired boilers with gas or electric heaters in homes. Local officials were held accountable through specific environmental targets. The results were remarkable: Beijing recorded 288 days of good air quality in 2021, a significant improvement from just 176 days in 2013

(Associated Press, 2022). This progress demonstrated how international events could serve as powerful drivers for lasting environmental reform, rather than just temporary solutions.

The first Olympic experience served as a catalyst for more comprehensive and long-term strategies to address air pollution not just in the host city, Beijing, but nationwide. In 2013, China introduced the Air Pollution Prevention and Control Action Plan, which set specific targets for reducing PM2.5 levels in key regions (Yu *et al.*, 2022). This was followed by the more comprehensive Environmental Protection Law in 2015, which significantly increased penalties for polluters and empowered environmental authorities (Wang *et al.*, 2023). These legislative measures marked a shift toward a more robust regulatory framework for environmental protection.

One of the most significant aspects of China's air pollution control strategy has been its focus on industrial restructuring and technological upgrades. The government initiated programs to phase out outdated production capacities in heavy industries such as steel, cement, and coal-fired power plants. Simultaneously, it promoted the adoption of cleaner technologies and stricter emission standards. The case of Shanxi Province, a major coal-producing region, illustrates this approach. Shanxi implemented a comprehensive plan to reduce its reliance on coal and diversify its economy by exploring solar, wind, hydrogen, and geothermal energy. This also included closing small, inefficient mines, investing in renewable energy projects, and promoting the development of high-tech industries. Shanxi achieved a 10.9 percent reduction in energy consumption from 2021 to 2023 (Yong and Mingli, 2024). Moreover, the coal mining process, which historically heavily relied on manual labor, is now substantially mechanized, with robots performing inspections and testing-related activities. Shanxi had 133 intelligent coal mines in operation by 2024 (Yong and Mingli, 2024). This transformation not only contributed to reduced air pollution but also facilitated economic restructuring toward more sustainable industries.

Energy mix optimization has been another crucial component of China's air pollution control strategy. The country has made significant strides in diversifying its energy sources, becoming the world's largest and fastest growing producer of renewable energy. Between 2010 and 2020, China increased its installed capacity of wind and solar power markedly, and in 2020, it deployed 72 GW of wind energy and 48 GW of solar capacity, vastly more than expected (Barnard, 2021) More interestingly, during 2021–2022, "the average annual increase in China's wind capacity was 178.6 terawatt hours (TWh), or 350% more than the average annual increase from 2015 through 2020" (Maguire, 2023). This rapid expansion of renewable energy capacity has been complemented by increased use of natural gas as a cleaner alternative to coal, particularly in urban areas. China has also made strides in its nuclear energy capacity. In 2018, the Haiyang Nuclear Power Plant began cogeneration of both heat and power. The plant expanded its capacity from 31.5 megawatts (MW) in 2019 to 202.5 MW in 2021 to 365 MW in 2022 (Van Wyk, 2022). The coastal city of Haiyang in Shandong Province of eastern China earned the title of "China's first zero-carbon heating city" in 2021.

The city of Xi'an provides a compelling case study of energy transition at the municipal level. Between 2002 and 2004, Xi'an undertook an ambitious project to replace coal-fired boilers with natural gas and electric boilers in urban areas (*China Daily*, 2004). After the State Council passed the Air Pollution Prevention and Control Action Plan in 2013, there was a broader "coal-to-gas" and "coal-to-electricity" campaign implemented across northern China. As described by Wang *et al.* (2020), since 2017, the government has taken action in Beijing and "the "2+26" cities located along the air pollution transport channel of the Beijing–Tianjin–Hebei (BTH) region" to replace "traditional household coal-fired stoves with wall-mounted natural gas heaters ("coal-to-gas") or electric stoves ("coal-to-electricity")" (p. 31018). Xi'an also undertook similar efforts, and in 2020, replaced its remaining coal-fired boilers with more environmentally friendly alternatives (*Xi'an Net News*, 2020). The project not only significantly reduced PM2.5 emissions but also improved energy efficiency in residential heating.

However, the transition has not been without hiccups. When the government replaced coal-fired boilers with a clean natural gas furnace, it resulted in higher heating bills for many residents, thereby putting a serious strain on the finances of those with lower levels of income. DeVore (2019) detailed how a rural construction worker saw "his winter heating bill triple, now consuming 13.3% of his annual income," in addition to raising concerns about whether this would be sustainable when government subsidies expired in three years. On another occasion, many villagers had to go to sleep fully clothed when there was a delay in installing the natural gas furnaces after the coal-fired boilers were taken away. Some villagers admitted to surreptitiously burning coal at night to keep their families warm despite knowing that there would be penalties if caught by the authorities (DeVore, 2019).

The environmental transformation, while necessary, has often placed disproportionate burdens on society's most vulnerable members. Beyond the increased heating costs, many small factory workers lost their livelihoods when polluting industries were shut down with little warning or support for the transition. Street vendors saw their incomes dwindle when industrial areas were suddenly relocated. In some rural areas, farmers who had relied on coal-burning kilns for generations to dry their crops were left scrambling for alternatives, affecting their agricultural output and income. Yet, amidst these challenges, countless ordinary citizens have shown remarkable resilience and dignity. Like the locust tree that grows slowly but surely, spreading its branches to provide shade for future generations, these nameless individuals endure present hardships knowing their sacrifices will benefit their children and grandchildren. From the elderly grandmother who learned to use a new electric stove to the young factory worker who retrained for a job in renewable energy to the street vendor who adapted his business to new environmental regulations, these common people stand as quiet heroes in China's environmental transformation. They embody the ancient wisdom that one generation plants the trees under whose shade future generations will rest, their daily sacrifices contributing to a legacy of cleaner air and healthier communities.

Much like Xu Sanguan in Yu Hua's (1995/2003) *Chronicle of a Blood Merchant*, who repeatedly sold his blood to support his family through economic hardship, these Chinese residents often find themselves making difficult sacrifices for survival. Just as Xu carefully balanced his blood-selling intervals to maintain his health while meeting his family's needs, today's rural residents must weigh the trade-offs between financial stability and compliance with environmental regulations. The clandestine burning of coal at night mirrors Xu's desperate measures during the Cultural Revolution — both represent acts of quiet rebellion born not of defiance, but of necessity. And just as Xu's blood-selling took a hidden toll on his body, the financial strain of these environmental policies extracts an unseen cost from China's working class, revealing how even well-intentioned reforms can place the heaviest burden on those least able to bear it.

Transportation sector reforms have also played a crucial role in China's air pollution mitigation efforts. The country has implemented increasingly stringent vehicle emission standards, promoted electric vehicles, and invested heavily in public transportation infrastructure. The city of Shenzhen stands out as a pioneer in this regard. In 2017, Shenzhen electrified its public bus fleet, comprising over 16,000 buses (Li *et al.*, 2020). It now has more EVs than any other city in the world, including the largest e-bus and e-taxi fleets (Li *et al.*, 2020). This transition not only significantly reduced vehicular emissions but also served as a model for other Chinese cities. Shenzhen's electric bus fleet reduced CO_2 emissions dramatically compared to a diesel bus fleet, with concomitant reductions in particulate matter and nitrogen oxides.

Enhanced monitoring and enforcement mechanisms have been critical to the success of China's air pollution control efforts. The country has established a nationwide air quality monitoring network and has increased transparency by making real-time air quality data publicly available. This has improved accountability and facilitated more targeted interventions. For instance, the Beijing–Tianjin–Hebei region, one of the most polluted areas in China, has implemented a joint prevention and control mechanism that allows for coordinated responses to severe air pollution events across multiple jurisdictions. In addition, the Chinese government banned crypto mining in 2021 in order to curb high levels of pollution. This decision was in alignment with President Xi Jinping's pledge that China would reach net-zero carbon emissions by 2060. However, mining models show that some Chinese miners, instead of relocating, simply went underground and continued to operate illegally (Carreras, 2024).

China's approach to air pollution control represents a fascinating case study of how environmental governance can expand or constrain human capabilities (Sen, 1992). Amartya Sen argues that expanding people's capabilities, which refer to "the skills, abilities, and understanding of individuals with respect to particular choices" (Preston, 1984, p. 961), is central to freedom and consequently development. Sen's capability approach treats the freedom of individuals as the basic building block by contending that "the usefulness of wealth lies in the things that it allows us to do- the substantive freedoms it helps us to achieve" (Sen, 1999, p. 14). Unlike traditional development metrics that focus solely on economic growth, China's environmental transformation reflects a growing recognition that true development encompasses the freedom to breathe clean air and live in a healthy environment.

The country's journey from severe air pollution to improved air quality illustrates how environmental protection and human capabilities are inextricably linked. Consider the transformation in Beijing. In 2006, when sandstorms regularly engulfed the capital, residents faced severe restrictions on their daily activities. Children couldn't play outdoors, the elderly were confined to their homes, and athletes couldn't train effectively. These restrictions represented "unfreedoms" (Sen, 1999) — conditions that prevented people from realizing their full potential. By implementing comprehensive air quality improvements, Beijing has gradually restored these basic freedoms to its residents. The 288 days of good air quality in 2021 weren't just statistics; they represented expanded opportunities for millions to live more fulfilling lives.

China's environmental transformation has been marked by a unique interplay between state action and collective awareness. Unlike traditional social movements that emerge from grassroots activism, China's environmental consciousness has largely been shaped by top-down policies combined with growing public awareness of environmental rights. The government's "war on pollution" gained legitimacy not just through regulatory enforcement but also through a collective recognition that clean air is a fundamental right rather than a luxury. This shift in perspective has manifested in interesting ways. When residents in Xi'an faced the transition from coal to natural gas heating, their struggles weren't just about economic hardship but reflected a broader tension between different types of freedoms. The right to affordable heating competed with the right to clean air, illustrating how environmental policies can create new capabilities while temporarily constraining others.

While China's environmental transformation has expanded certain freedoms, it has also created new forms of constraint, particularly for vulnerable populations. The story of the rural construction worker whose heating bills tripled reveals how environmental progress can sometimes come at the cost of economic freedom for society's most vulnerable members. These tensions echo broader debates about how societies balance different types of capabilities — environmental, economic, and social. Yet, these challenges have also sparked innovative responses. Communities have developed new coping mechanisms, from collective purchasing of cleaner heating equipment to informal support networks helping elderly residents adapt to new technologies. These adaptations demonstrate how communities can enhance their collective capabilities even in the face of major societal transitions.

China's journey in mitigating air pollution demonstrates the effectiveness of a multi-faceted approach combining regulatory measures, technological innovation, and structural economic changes. The country's experience offers valuable lessons for other rapidly industrializing nations grappling with similar environmental challenges. However, as China's economy continues to grow, maintaining and improving air quality will require ongoing efforts and innovation. Future research should focus on the long-term sustainability of these improvements, the potential for replicating successful strategies in other contexts, and the integration of air pollution control with broader climate change mitigation efforts.

References

Associated Press. (2022, February 9). How China got blue skies in time for Olympics. *NBC News*. Retrieved on November 3, 2024 from https://www.nbcnews.com/science/environment/china-got-blue-skies-time-olympics-rcna15340.

Barnard, M. (2021). A decade of wind, solar, & nuclear In China shows clear scalability winners. *CleanTechnica*. Retrieved on November 3, 2024 from https://cleantechnica.com/2021/09/05/a-decade-of-wind-solar-nuclear-in-china-shows-clear-scalability-winners/.

Carreras, T. (2024, November 1). Bitcoin mining bans can backfire on climate conscious governments, a new research finds. *CoinDesk*. Retrieved on November 3, 2024 from https://www.coindesk.com/policy/2024/11/01/bitcoin-mining-bans-can-backfire-on-climate-conscious-governments-a-new-research-finds/.

China Daily. (2006, April 19). 330,000-ton sand fell on Beijing. Retrieved on November 3, 2024 from https://www.chinadaily.com.cn/china/2006-04/19/content_571196.htm.

China Daily. (2004, August 3). Xi'an cleans up coal boilers in urban areas. Retrieved on November 3, 2024 from http://www.china.org.cn/english/travel/102829.htm.

DeVore, C. (2019). It's cold in China, and environmental central planning has turned off the heat. *Forbes*. Retrieved on November 3, 2024 from https://www.forbes.com/sites/chuckdevore/2019/01/23/its-cold-in-china-and-environmental-central-planning-has-turned-off-the-heat/.

Hua, Y. (2003). *Chronicle of a Blood Merchant* (A.F. Jones, Trans.). New York: Anchor Books. (Original Work Published 1995).

Li, M., Ye, H., Liao, X., Ji, J., and Ma, X. (2020). How Shenzhen, China pioneered the widespread adoption of electric vehicles in a major city: Implications for global implementation. *Wiley Interdisciplinary Reviews: Energy and Environment*, 9(4), e373.

Liu, L. and Ogunc, A. (2023). Beijing Blue: Impact of the 2008 Olympic Games and 2014 APEC Summit on Air Quality. *Atlantic Economic Journal*, 51(1), 83–100.

Maguire, G. (2023, March 1). China widens renewable energy supply lead with wind power push. *Reuters*. Retrieved on November 3, 2024 from https://www.reuters.com/markets/commodities/china-widens-renewable-energy-supply-lead-with-wind-power-push-2023-03-01/.

Preston, L. M. (1984). Freedom, markets, and voluntary exchange. *American Political Science Review, 78*(4), 959–970.

Sen, A. (1992). *Inequality Reexamined*. Cambridge, MA: Harvard University Press.

Sen, A. (1999). *Development as Freedom*. Oxford: Oxford University Press.

Van Wyk, B. (2022, November 18). Build first, destroy later: China is slowly replacing dirty coal with clean nuclear heating. *The China Project*. Retrieved on November 3, 2024 from https://thechinaproject.com/2022/11/18/build-first-destroy-later-china-is-slowly-replacing-dirty-coal-with-clean-nuclear-heating/.

Wang, H., Li, T., Zhu, J., Jian, Y., Wang, Z., and Wang, Z. (2023). China's new environmental protection law: Implications for mineral resource policy, environmental precaution and green finance. *Resources Policy, 85*, 104045.

Wang, S., Su, H., Chen, C., Tao, W., Streets, D. G., Lu, Z., Zheng, B., Carmichael, G. R., Lelieveld, J., Pöschl, U., and Cheng, Y. (2020). Natural gas shortages during the "coal-to-gas" transition in China have caused a large redistribution of air pollution in winter 2017. *Proceedings of the National Academy of Sciences, 117*(49), 31018–31025.

Xi'an Net News. (2020, August 3). In 2020, Xi'an will stop 39 coal-fired boilers with 2,855 steam tons. Retrieved on November 3, 2024 from https://m.xinli-boiler.com/news/industry-news/85.html.

Yong, H. and Mingli, F. (2024, September 25). Coal-rich province Shanxi makes progress in energy revolution. *People's Daily Online*. Retrieved on November 3, 2024 from http://en.people.cn.

Yu, Y., Dai, C., Wei, Y., Ren, H., and Zhou, J. (2022). Air pollution prevention and control action plan substantially reduced PM2.5 concentration in China. *Energy Economics, 113*, 106206.

Chapter 7

India's Vanishing Blue Skies: A Tale of Growth and Neglect

On January 29th 2023, Mumbai, India's financial capital and traditionally a city with better air quality thanks to its coastal location, experienced an unprecedented pollution crisis that shocked its residents and experts alike. The city's Air Quality Index hit 325 — a level categorized as "hazardous" on the international AQI scale used by the US Environmental Protection Agency and labeled as "very poor" on India's national scale (Tembhekar, 2023). This reading was severe enough to surpass even Delhi's pollution levels that day, a virtually unheard of occurrence that sent alarming signals about the spreading nature of India's air quality crisis. The reason lay in a confluence of factors — ongoing construction projects, vehicular emissions, industrial activity, and unusually stable atmospheric conditions that trapped pollutants close to the ground. As a thick haze shrouded the city's iconic skyline, including the historic Gateway of India and the gleaming towers of Nariman Point, visibility dropped dramatically. Mumbai's residents, unaccustomed to such severe air pollution, found themselves suddenly grappling with burning eyes, respiratory difficulties, and the need to wear masks — not for COVID-19, but to filter the toxic air. This extraordinary event in a coastal city known for its cleansing sea breezes served as a stark warning that India's air pollution crisis was no longer confined to its northern regions but was expanding its deadly reach across the nation.

India's rapid economic ascension in the 21st century has been accompanied by a parallel narrative of environmental degradation, most visibly manifested in the deterioration of air quality across its burgeoning urban landscapes. The trajectory of India's economic growth has been remarkable, with the country being ranked as the fifth-largest economy in the world in 2024, ahead of the United Kingdom (World Population Review, 2024). Moreover, according to a report by S&P Global, India is set to become the third-largest economy in the world by 2030–2031 if the predicted annual GDP growth rate of 6.7% is realized (Joshi *et al.*, 2024). This growth has lifted millions out of poverty and positioned India as a major player in the global economy. The transformation has been particularly visible in sectors like information technology, pharmaceuticals, and manufacturing, which have created new employment opportunities but also contributed to urban air pollution through increased energy consumption and industrial emissions. As such, this progress has come at a significant environmental cost, particularly in terms of air quality. In 2023, a shocking 35 of the world's top 40 most polluted cities were located in India, with Begusarai ranking the highest (IQAir, 2024). This statistic becomes even more alarming when considering that these cities represent diverse geographical regions and economic profiles, suggesting that air pollution has become a systemic rather than localized problem.

The scale of the air pollution problem in India is staggering. A study published by de Bont *et al.* (2024) on 10 cities in India between 2008 and 2019 found that 7.2% of all daily deaths could be attributed to PM2.5 concentrations higher than the WHO guidelines. In another study, India State-Level Disease Burden Initiative Air Pollution Collaborators (2021) noted that 1.67 million deaths were attributable to air pollution in India in 2019, accounting for about 17.8% of the total deaths in the country. The authors also estimated that in 2019, India lost $36.8 billion due to air pollution's impact on human health — $28.8 billion because people died too early and couldn't contribute to the economy and another $8 billion because sick workers had to miss work or couldn't perform at their best due to pollution-related illnesses (India State-Level Disease Burden Initiative Air Pollution Collaborators, 2021). These economic losses stand out in the context of India's development aspirations, as they represent resources that could have been invested in infrastructure, healthcare, or education.

Just four decades ago, life in India moved at a different pace. Cities and towns were marked by the gentle rhythm of bicycle bells and the steady clip-clop of bullock carts. Letters took days to reach their destinations, and people walked or cycled to nearby markets. The streets of Old Delhi, now choked with traffic and smog, were once filled with the aroma of street food and the chatter of locals gathering in community spaces.

Seasonal changes were marked by clear visual cues — the crisp visibility of winter mornings, the vibrant colors of spring flowers, and the dramatic cloud formations of approaching monsoons. While this slower life had its challenges, it came with an invaluable benefit that most Indians took for granted — clean air. The skyline was clear and the hills visible from the city edges weren't shrouded in perpetual haze. Children played outdoors without the constant threat of respiratory illness and the elderly didn't need air purifiers to breathe safely in their own homes. Today, in the rush toward modernization, millions of Indians find themselves trapped in a cruel paradox — while their standard of living has improved, their quality of life has deteriorated. The toxic air has become a silent, persistent killer, gradually weakening lungs, straining hearts, and shortening lives across all age groups and social classes. It is a price paid not just in statistics and economic losses, but in countless personal tragedies — the aspiring athlete who can no longer run, the elderly who are confined indoors, and the children who will never know what truly clean air feels like.

Interestingly, urbanization and industrialization have been key drivers of both economic growth and air pollution in India. Dutta *et al.* (2021) stated that "rapid and uncontrolled urbanization coupled with population growth, rising vehicle population, and growth of industries" (p. 93) contributed to the escalating problem of pollution. The automotive sector exemplifies this dual impact — while it has been a significant contributor to India's economic growth, employing millions and attracting foreign investment, it has also led to a massive increase in vehicular emissions. According to the records of the Indian *Ministry of Road Transport and Highways*, by 2022, India had over 350 million registered vehicles, compared to around only 300,000 in 1951. This situation is also echoed in the visible transformation of India's cityscapes, where construction dust, vehicular emissions, and industrial pollutants have become pervasive. The case of Delhi, India's capital, exemplifies the severity of the air quality crisis. During winter months, the city frequently experiences "severe" air quality conditions, with PM2.5 levels exceeding 300 $\mu g/m^3$, far above the WHO guideline of 10 $\mu g/m^3$. According to the Chief Minister of Delhi, living in Delhi in these times is like being in a gas chamber and breathing is equivalent to smoking 50 cigarettes a day (Basu, 2021).

Despite recognition of the problem and various policy initiatives over a long time, measures to curb air pollution in India have not yielded the expected results. Several factors contribute to this inefficacy. The fragmented governance structure, where air pollution control falls under the purview of multiple agencies at the central, state, and local levels, leads to coordination challenges and diluted accountability. For instance, while the Central Pollution Control Board under the *Ministry of Environment, Forest and Climate Change* sets standards and guidelines, state pollution control boards are responsible for implementation, and urban local bodies handle day-to-day monitoring and enforcement. This multi-layered structure often results in delayed responses to pollution emergencies and inconsistent application of regulations. As National Green Tribunal Chairperson Swatanter Kumar observed years ago, "No one is responsible for any pollution, it seems. Everyone is justifying their actions" (*Hindustan Times*, 2015).

Inadequate enforcement of existing regulations further exacerbates the problem. While India has established ambient air quality standards and emission norms, their implementation remains weak. For example, the installation of the mandated flue gas desulfurization technology (which helps in controlling sulfur dioxide emissions) in India's coal-fired thermal plants has met with various hurdles. This included limited availability of vendors with the capacity to supply and install relevant components, problems with importing the required technology, equipment, and skilled manpower from other countries, and issues with standardization due to varying needs like space constraints (Ministry of Power, 2024). The challenges extend beyond technical issues to include regulatory loopholes, insufficient penalties for non-compliance, and a lack of transparent monitoring systems that would allow public oversight of pollution levels and enforcement actions.

Limited capacity at the local level also hampers effective air quality management. Many urban local bodies lack the technical expertise and financial resources to implement and monitor air quality improvement measures effectively. This data gap impedes the development of targeted, evidence-based interventions. For example, many cities lack sufficient air quality monitoring stations, leading to incomplete data about pollution patterns and sources. Those that do have monitoring systems often struggle with maintenance and calibration issues, resulting in unreliable data. Socio-economic constraints further complicate efforts to reduce pollution, as these often conflict with short-term economic interests and livelihoods. For instance, India banned the burning of agricultural residues in 2015, but the practice continued unabated because burning is the cheapest alternative for farmers. This highlights the need for holistic approaches that address both environmental and economic concerns. The government's efforts to provide alternatives, such as subsidized machinery for crop residue management, have had limited success due to high operational costs and the short window available between harvesting one crop and sowing the next.

The contrast between India's struggle with air pollution and China's relative success in recent years is noteworthy. While both countries faced similar challenges of rapid industrialization and urbanization, China has made significant strides in improving air quality, particularly in its major cities. Several factors explain why India has been unable to replicate China's success. China's centralized governance structure allows for swift implementation of pollution control measures across regions. In contrast, India's quasi-federal system and diverse political landscape often lead to policy inconsistencies and implementation delays. Moreover, China's "war on pollution" received substantial political and financial backing. India, still grappling with significant poverty and development challenges, has been more constrained in allocating resources to environmental issues. Through their analysis of city-level panel data in India from 1986 to 2007, Greenstone and Hanna (2014) showed that air pollution regulations can be quite effective at reducing ambient concentrations of particulate matter, though such efforts need widespread political support from within.

Technological adoption has also played a crucial role in China's air quality improvements. China has rapidly embraced clean technologies, becoming the world's largest investor in renewable energy. The country has also taken aggressive steps to electrify its transportation sector, with electric vehicles accounting for a significant portion of new car sales. India, while making progress, has been slower in transitioning its energy sector and industrial processes. This disparity in technological uptake has significant implications for the pace of air quality improvement. Public awareness and pressure have been another differentiating factor. Public concern about air pollution in China reached a tipping point, creating strong domestic pressure for action. While awareness is growing in India, it has not yet translated into the same level of sustained public demand for change. This underscores the importance of public engagement and environmental education in driving policy action. The success of China's air quality improvement efforts also highlights the value of comprehensive planning and sustained commitment to implementation, aspects that India could potentially adapt to its own context while accounting for its unique political and social structures.

India's journey toward cleaner air is intrinsically linked to its broader development narrative. The country faces the formidable challenge of balancing economic growth with environmental sustainability. Dr. Arvind Kumar, chest surgeon and founder of the Lung Care Foundation, distressingly stated the following (Whiting, 2022):

> When I operate, even on children, I see black deposits. There is a sea change in the demography of lung cancer: from smoking, it is moving more and more towards air pollution. This is a very serious development: from the pink lungs that we are born with, due to the impact of pollution, they become black. And as a chest surgeon, I see them everyday when I operate on these lungs.

The medical community has been increasingly vocal about the health impacts of air pollution, with doctors reporting a rise in respiratory ailments even among non-smokers and young children. Addressing India's air pollution crisis will require a paradigm shift in governance, stronger enforcement mechanisms, substantial investments in clean technologies, and a reimagining of urban development patterns. Only through such comprehensive and sustained efforts can India hope to reclaim its blue skies while continuing on its path of economic progress.

The air pollution crisis in India represents more than just an environmental challenge — it embodies what development economists call a fundamental "unfreedom" that systematically constrains human capabilities and potential (Sen, 1999). This paradox illuminates a crucial aspect of India's development narrative — the expansion of one set of freedoms (economic opportunities) has led to the contraction of another (environmental well-being) that traps individuals and communities in cycles of disadvantage. The solution requires rethinking development itself, moving beyond economic metrics to consider what Sen (1999) calls "development as freedom" — the enhancement of human capabilities in all their dimensions, including the ability to live in a healthy environment.

The path forward for India necessitates a multi-faceted approach that addresses the root causes of air pollution while promoting sustainable economic development. This may involve reimagining urban planning to prioritize green spaces and public transportation, incentivizing clean energy adoption across industries, and fostering a culture of environmental stewardship among citizens. Furthermore, strengthening institutional capacity for environmental governance and promoting regional cooperation on air quality management will be crucial in tackling this transboundary issue. As India strives to balance its aspirations for economic growth with the imperative of environmental protection, the resolution of its air pollution crisis will serve as a litmus test for its commitment to sustainable development. The country's ability to navigate this challenge will not only determine the health and well-being of its citizens but also shape its role in global efforts to combat climate change and environmental degradation.

References

Basu, M. (2019). The great smog of Delhi. *Lung India, 36*(3), 239–240.

de Bont, J., Krishna, B., Stafoggia, M., Banerjee, T., Dholakia, H., Garg, A., Ingole, V., Jaganathan, S., Kloog, I., Lane, K., Mall, R., Mandal, S., Nori-Sarma, A., Prabhakaran, D., Rajiva, A., Tiwari, A., Wei, Y., Wellenius, G., Schwartz, J., Prabhakaran, P., and Ljungman, P. (2024). Ambient air pollution and daily mortality in ten cities of India: A causal modeling study. *The Lancet Planetary Health, 8*(7), e433–e440.

Dutta, S., Ghosh, S., and Dinda, S. (2021). Urban air-quality assessment and inferring the association between different factors: A comparative study among Delhi, Kolkata and Chennai megacity of India. *Aerosol Science and Engineering, 5*, 93–111.

Greenstone, M. and Hanna, R. (2014). Environmental regulations, air and water pollution, and infant mortality in India. *American Economic Review, 104*(10), 3038–3072.

Hindustan Times. (2015, July 14). NGT raps road transport ministry over Delhi pollution inaction. Retrieved on November 14, 2024 from https://www.hindustantimes.com.

India State-Level Disease Burden Initiative Air Pollution Collaborators. (2021). Health and economic impact of air pollution in the states of India: The global burden of disease study 2019. *The Lancet Planetary Health, 5*(1), e25–e38.

IQAir. (2024). World's most polluted cities: Historical rankings 2017-2023. Retrieved on November 14, 2024 from https://www.iqair.com/us/world-most-polluted-cities.

Joshi, D., Luchnikava-Schorsch, H., De Lima, P., and Rana, V. (2024, September 19). India's growing role in the global economy. *S&P Global.* Retrieved on November 14, 2024 from https://www.spglobal.com/en/research-insights/special-reports/india-forward/indias-growing-role-in-the-global-economy.

Ministry of Power. (2024, August 1). Status of flue gas de-sulphurisation (FGD) installation in thermal power plants. *PIB Delhi.* Retrieved on November 14, 2024 from https://pib.gov.in/PressReleaseIframePage.aspx?PRID=2040090.

Sen, A. (1999). *Development as Freedom.* Oxford: Oxford University Press.

Tembhekar, C. (2023, January 29). Mumbai's AQI dips again; could be worst January in five years. *Times of India.* Retrieved on October 24, 2023 from https://timesofindia.indiatimes.com/city/mumbai/mumbais-aqi-dips-again-could-be-worst-january-in-five-years/articleshow/97408021.cms.

Whiting, K. (2022, December 8). This is what air pollution is doing to teenagers' lungs, according to a chest surgeon. *World Economic Forum.* Retrieved on October 24, 2023 from https://www.weforum.org/stories/2022/12/air-pollution-lungs-children-india/.

World Population Review. (2024). GDP Ranked by Country 2024. Retrieved on November 14, 2024 from https://worldpopulationreview.com/countries/by-gdp.

Chapter 8

Clearer Horizons Forward: Developing Nations' Solutions and Challenges

In the sprawling mining complexes of Inner Mongolia, China, a quiet revolution is taking place. The Baiyun Obo Mining District, one of the world's largest rare earth mining operations, has transformed into what Chinese officials call an "intelligent mine" (Wang *et al.*, 2022). Since 2021, the mine has deployed autonomous trucks, smart ventilation systems, and AI-powered environmental monitoring stations that work in concert to reduce emissions while maintaining productivity. The intelligent system acts like a living organism, automatically adjusting ventilation based on real-time air quality data, predicting potential pollution hotspots before they form, and rerouting mining operations to minimize environmental impact — thereby significantly reducing energy consumption compared to traditional mining operations while dramatically decreasing dust and pollutant emissions.

China realized very early the critical role technology can play in combating pollution, and "the number of federal monitoring stations across China nearly tripled between 2012 and 2020, from 661 to 1,800" (Jain and Kolla, 2023). However, technological innovation to combat pollution is not isolated to China. Across the developing world, countries are endeavoring to leverage cutting-edge technology to combat pollution. In India's capital region, the Delhi government's "Real-time Source Apportionment" project (IIT Kanpur *et al.*, 2023), launched in 2023, uses advanced AI algorithms and a dense network of sensors to identify pollution sources in real time. Meanwhile, in Indonesia, authorities have partnered with

Google's Environmental Insights Explorer (EIE) to deploy sophisticated machine learning models that analyze satellite data and traffic patterns to generate estimates of greenhouse gas emissions from buildings and transportation. In April 2022, West Nusa Tenggara (NTB) in Indonesia became EIE's first launch location in Southeast Asia (Google Indonesia, 2022).

The road ahead for developing countries in managing air pollution and mitigating its deleterious effects presents a complex tapestry of challenges and opportunities. These nations find themselves at a critical juncture where the imperative of economic growth intersects with the urgent need for environmental stewardship, particularly in the realm of air quality management. The scale of the air pollution crisis in these developing nations is staggering. According to the World Health Organization (WHO), 99% of the global population breathes air that exceeds WHO guideline limits (WHO, 2022), with low- and middle-income countries suffering from the highest exposures. The health implications are profound, particularly for vulnerable populations such as children, with air pollution responsible for 572,000 neonatal deaths globally in 2021, representing 26% of the total newborn deaths (Health Effects Institute, 2024).

One critical area of focus is the energy sector. Many developing countries rely heavily on fossil fuels, particularly coal, for power generation, a dependency rooted in historical industrialization patterns and economic constraints. This reliance stems from coal's historical role as a readily available, cost-effective fuel source that enabled rapid industrial growth in the 20th century. The established infrastructure, from mining operations to power plants, represents massive sunk costs that create economic and technical barriers to transition. Furthermore, coal mining and related industries often serve as significant employers in developing regions, making the shift away from coal not just an energy challenge, but a complex socio-economic issue involving workers' livelihoods and regional economic stability. For instance, in India's coal belt of Jharkhand, which supports nearly 300,000 direct coal mining jobs, making any kind of "just" transition is particularly challenging. While countries like India have set ambitious renewable energy targets, aiming for 450 GW of renewable energy capacity by 2030 (PIB Delhi, 2021), the transition faces substantial hurdles. These include the high initial capital costs of renewable infrastructure, the challenge of retiring economically viable but polluting coal plants before the end of their operational life, and the need to ensure energy security and affordability for growing populations.

The intermittent nature of renewable energy sources also requires significant grid modernization investments, which many developing nations struggle to finance.

This complex energy landscape intersects directly with urban planning and transportation challenges, as cities in developing nations grapple with both power demands and mobility needs. Rapid urbanization in countries like Indonesia and Pakistan has led to sprawling cities with inadequate public transportation systems, resulting in high levels of vehicular emissions. The situation is particularly acute in megacities like Jakarta, where traffic congestion costs the economy estimated billions of dollars while contributing significantly to air pollution. Innovative approaches to urban development, such as transit-oriented design and the promotion of non-motorized transport, can help alleviate this issue. For instance, Bogotá's *TransMilenio* bus rapid transit system demonstrates the potential for cost-effective, low-emission urban transport solutions (Kimmelman, 2023). In Peru, Lima's *Respira Limpio* campaign is striving to reduce emissions by identifying vehicles with high emissions and facilitating their repair (C40 Cities, 2023). Similarly, Bengaluru's *TenderSure Project* has been working on transforming roads into more walkable and bike-friendly streets, with properly designed footpaths, parking bays, better storm drainage systems, and proper landscaping (C40 Cities, 2023).

Institutional capacity building and governance reform are critical components of any comprehensive air quality management strategy. Many developing countries lack robust air quality monitoring networks and enforcement mechanisms. Strengthening these systems is essential for evidence-based policymaking and effective implementation of pollution control measures. The establishment of India's National Clean Air Programme (NCAP) in 2019 represents a step in this direction, setting time-bound targets for air pollution reduction across 122 cities (Ganguly *et al.*, 2020). The program has already shown initial results, with several cities reporting some improvements in air quality metrics. However, the efficacy of such programs hinges on consistent political will and adequate resource allocation. South Africa exemplifies this approach, where the South African Consortium of Air Quality Monitoring created a low-cost air quality monitoring system based on sensors, the Internet of Things, and Artificial Intelligence called AI_r (Mellado, 2024). AI_r does not require people to collect air quality samples, but rather automatically measures the concentration of PM2.5 and uploads the data to the cloud in real time. As such, the system can help to effectively identify pollution hotspots, instead of just monitoring the average air quality in the locality.

Education and public awareness are fundamental to creating sustained demand for clean air policies. Developing countries need to invest in environmental education at all levels, from schools to community organizations. Citizen science initiatives, such as low-cost air quality monitoring projects, can empower communities to advocate for their right to clean air. Taiwan, for example, has witnessed the emergence of the AirBox project, where residents use low-cost sensors to monitor air quality and provide actionable data about their environments (Mahajan *et al.*, 2021). The success of grassroots movements in driving policy change offers valuable lessons for other developing nations. These initiatives demonstrate how informed citizens can become powerful catalysts for environmental change, transforming abstract policy goals into tangible community action.

The rapid advancement of artificial intelligence (AI), generative AI, and robotics presents both opportunities and challenges in the context of global pollution displacement and environmental sustainability. One of the most promising applications of AI in addressing global pollution is in the realm of environmental monitoring and prediction. Advanced machine learning algorithms can process vast amounts of data from satellites, ground sensors, and other sources to provide more accurate and timely information about air quality, water pollution, and other environmental indicators. Deep learning models can be utilized to estimate air pollution levels from satellite imagery with high accuracy. This capability could significantly enhance our ability to track pollution displacement across borders and hold nations accountable for their environmental impacts. Moreover, AI-powered predictive models can forecast pollution trends and identify potential hotspots before they become critical. This predictive capability could enable more proactive and targeted interventions, potentially mitigating the environmental costs of industrial activities in developing countries.

Generative AI, a subset of AI that can create new content, has the potential to revolutionize product design and manufacturing processes. By optimizing designs for efficiency and sustainability, generative AI could help reduce the environmental footprint of mass production. Generative design algorithms could create more efficient structural components, potentially reducing material use and energy consumption in manufacturing. Furthermore, generative AI could contribute to the development of new materials and processes that are inherently more environmentally friendly. As pointed out by Papadimitriou *et al.* (2024), AI-assisted materials discovery and design could potentially accelerate the development of new types of sustainable materials, thereby reducing the pollution associated with traditional manufacturing processes. This could help address the environmental challenges associated with production in both developed and developing countries.

The advancement of robotics and automation technologies could potentially alter the global distribution of manufacturing activities. As robots become more sophisticated and cost effective, the economic incentives for offshoring production to countries with cheaper labor may diminish. This phenomenon, often referred to as "reshoring," could lead to a redistribution of manufacturing activities and, consequently, a shift in global pollution patterns. While this trend could potentially reduce pollution in some developing countries, it also raises questions about the economic implications for nations that have relied on manufacturing for economic growth and poverty reduction.

It is important to note that the environmental impact of reshoring is not straightforward. Fratocchi and Di Stefano (2019) observed that the environmental benefits of reshoring depend on the relative efficiency of production technologies and the stringency of environmental regulations in different countries. Thus, the net environmental impact of robotics-driven reshoring would depend on how these technologies are implemented and regulated across different national contexts. A further complexity arises from the energy intensity of automated manufacturing — while robots may operate with greater precision and less waste, they require significant electrical power, making the carbon footprint of production highly dependent on the energy mix of the host country. Additionally, reshoring could potentially lead to a "two-speed" manufacturing world, where developing countries that lack the capital or technical capacity to adopt advanced automation technologies risk being left behind in the global economy. This could create new environmental challenges as these nations might be forced to compete on cost by relaxing environmental standards or focusing on more polluting industries that are less amenable to automation.

Going forward, AI and big data analytics will inevitably play a crucial role in environmental policymaking and governance by processing complex datasets to identify policy interventions that optimize for both economic and environmental outcomes. AI could help design more effective and politically feasible climate policies by simulating the impacts of different policy scenarios. Moreover, AI could enhance the enforcement of environmental regulations by improving the detection of non-compliance. A study by Solórzano *et al.* (2023) demonstrated how spatio-temporal deep learning algorithms could be used to detect deforestation using remote sensing data, which in turn can potentially be utilized to identify illegal mining or construction activities. Similar approaches could be applied to monitor industrial activities and their associated emissions.

Beyond AI and robotics, a new wave of technologies promises to revolutionize pollution control and environmental protection efforts. Quantum sensors, leveraging the principles of quantum mechanics, are enabling unprecedented precision in detecting and measuring pollutants. Choy (2014) notes that "quantum sensors can detect hazardous materials in industrial settings in real time. This capability not only helps prevent accidents and protects workers from exposure to dangerous chemicals but also enables quick response times, mitigating the impact of pollution on the environment." In addition, biotechnology and synthetic biology are emerging as powerful tools in the fight against pollution. Engineered microorganisms capable of breaking down pollutants or converting them into harmless substances are showing promising results in initial trials (Borchert *et al.*, 2021; Kurade *et al.*, 2021). Moreover, advanced materials science is producing breakthrough solutions for pollution control. Metamaterials with engineered properties at the nanoscale are being researched to capture and neutralize pollutants more effectively than traditional filters. Smart materials that could adapt their properties based on environmental conditions might be particularly promising. Lastly, the integration of blockchain technology with environmental monitoring systems is creating new possibilities for transparency and accountability in pollution control. Smart contracts and distributed ledger technology could enable automated enforcement of environmental regulations and create relatively tamper-proof records of emissions and compliance.

While these technological developments offer significant potential for addressing global pollution challenges, they also raise important ethical and societal questions. Their implementation in environmental management and industrial production could exacerbate existing inequalities between developed and developing nations. The digital divide remains a significant barrier, with many developing countries lacking the infrastructure and expertise needed to fully leverage these technologies. Furthermore, the reliance on AI for environmental decision-making raises questions about transparency, accountability, and the role of human judgment in addressing complex socio-environmental issues. As we navigate this technological frontier, it is crucial to adopt an interdisciplinary approach that considers not only the technical capabilities of these technologies but also their broader societal and ethical implications.

Looking 50 years into the future, the landscape of pollution control and environmental protection could be transformed by technologies that today exist only in research laboratories or theoretical frameworks. Cities of 2075 might employ vast networks of bio-synthetic trees — engineered organisms that combine the natural ability of plants to absorb carbon dioxide with enhanced capabilities to process other pollutants. These artificial ecosystems could be integrated into urban infrastructure, creating self-regulating environmental control systems that adapt to changing pollution levels and weather patterns. Our future transportation systems could be revolutionized by magnetic levitation or gravity manipulation technology, enabling friction-free movement that produces zero emissions. In fact, even manufacturing could evolve to utilize molecular assembly technologies, where products are built atom by atom in closed systems that produce no waste or emissions. This technology, combined with advanced AI systems, could enable a truly circular economy where the concept of waste becomes obsolete. Similarly, environmental monitoring could advance to include sentient environmental systems — networks of advanced AI, quantum sensors, and bio-engineered organisms that work together to maintain ecological balance. These systems could predict and prevent pollution events before they occur, automatically adjusting industrial processes and urban systems to maintain optimal environmental conditions.

References

Borchert, E., Hammerschmidt, K., Hentschel, U., and Deines, P. (2021). Enhancing microbial pollutant degradation by integrating eco-evolutionary principles with environmental biotechnology. *Trends in Microbiology*, *29*(10), 908–918.

C40 Cities. (2023, May 2). Green News Digest: Cities tackling air pollution for a healthier future. *C40 Cities News & Insights*. Retrieved on January 3, 2024 from https://www.c40.org/news/green-news-digest-cities-tackling-air-pollution-healthier-future.

Choy, A. (2024, March 14). Unleashing quantum technology for next-generation environmental monitoring. *Verdantix*. Retrieved on April 5, 2024 from https://www.verdantix.com/insights/blogs/unleashing-quantum-technology-for-next-generation-environmental-monitoring.

Fratocchi, L. and Di Stefano, C. (2019). Does sustainability matter for reshoring strategies? A literature review. *Journal of Global Operations and Strategic Sourcing, 12*(3), 449–476.

Ganguly, T., Selvaraj, K. L., and Guttikunda, S. K. (2020). National Clean Air Programme (NCAP) for Indian cities: Review and outlook of clean air action plans. *Atmospheric Environment: X, 8*, 100096.

Google Indonesia. (2022, April 21). Google launches Environmental Insights Explorer in West Nusa Tenggara — first location in Southeast Asia. Retrieved on January 3, 2024 from https://blog.google/intl/id-id/company-news/technology/2022_04_environmental-insights-explorer-ntb/?.

Health Effects Institute. (2024). *State of Global Air 2024.* Special Report. Boston, MA: Health Effects Institute. Retrieved on November 5, 2024 from https://www.stateofglobalair.org/resources/report/state-global-air-report-2024.

IIT Kanpur, IIT Delhi, TERI, and Airshed. (2023). Real-time source apportionment and forecasting for advance air pollution management in Delhi. *Winter season report submitted to Delhi Pollution Control Committee — Delhi Government.*

Jain, A. and Kolla, S. (2023, March 15). Real time mapping is helping Delhi clear up its air. *Democracy News Live.* Retrieved on January 3, 2024 from https://democracynewslive.com/environment/real-time-mapping-to-clear-delhis-air-quality-1204854.

Kimmelman, M. (2023, December 7). How one city tried to solve gridlock for us all. *The New York Times.* Retrieved on January 3, 2024 from https://www.nytimes.com/interactive/2023/12/07/headway/bogota-bus-system-transmilenio.html.

Kurade, M. B., Ha, Y. H., Xiong, J. Q., Govindwar, S. P., Jang, M., and Jeon, B. H. (2021). Phytoremediation as a green biotechnology tool for emerging environmental pollution: A step forward towards sustainable rehabilitation of the environment. *Chemical Engineering Journal, 415*, 129040.

Mahajan, S., Luo, C. H., Wu, D. Y., and Chen, L. J. (2021). From Do-It-Yourself (DIY) to Do-It-Together (DIT): Reflections on designing a citizen-driven air quality monitoring framework in Taiwan. *Sustainable Cities and Society, 66*, 102628.

Mellado, B. (2024, August 14). Air pollution in South Africa: Affordable new devices use AI to monitor hotspots in real time. *The Conversation.* Retrieved on November 5, 2024 from https://theconversation.com/air-pollution-in-south-africa-affordable-new-devices-use-ai-to-monitor-hotspots-in-real-time-235897.

Papadimitriou, I., Gialampoukidis, I., Vrochidis, S., and Kompatsiaris, I. (2024). AI methods in materials design, discovery and manufacturing: A review. *Computational Materials Science, 235*, 112793.

PIB Delhi. (2021, October 11). India set to achieve 450 GW renewable energy installed capacity by 2030: Ministry of New and Renewable Energy. Press Information Bureau, Government of India. Retrieved on January 3, 2024 from https://pib.gov.in/Pressreleaseshare.aspx?PRID=1762960.

Solórzano, J. V., Mas, J. F., Gallardo-Cruz, J. A., Gao, Y., and de Oca, A. F. M. (2023). Deforestation detection using a spatio-temporal deep learning approach with synthetic aperture radar and multispectral images. *ISPRS Journal of Photogrammetry and Remote Sensing, 199*, 87–101.

Wang, G., Ren, H., Zhao, G., Zhang, D., Wen, Z., Meng, L., and Gong, S. (2022). Research and practice of intelligent coal mine technology systems in China. *International Journal of Coal Science & Technology, 9*(1), 24.

WHO. (2022, April 4). Billions of people still breathe unhealthy air: New WHO data. World Health Organization. Retrieved on January 3, 2024 from https://www.who.int/news/item/04-04-2022-billions-of-people-still-breathe-unhealthy-air-new-who-data.

Chapter 9

Tales of Caution: Green Gentrification and the Hidden Cost of Clean Air

On a crisp autumn morning in 2010, longtime residents of Greenpoint, Brooklyn, gathered to celebrate what seemed like a victory. Attorney General Andrew Cuomo announced that ExxonMobil will pay $25 million to compensate for the cleanup of its estimated 17 million gallons of oil spill since the mid-1900s and any related environmental contamination. This spill, one of the largest recorded in U.S. history, had seeped into the ground over decades, creating an underground plume that contaminated soil and groundwater across a huge area. Cuomo stated, "For far too long, residents of Greenpoint have been forced to live with an environmental nightmare lurking just beneath their homes, their businesses, and their community" (Kherani, 2010). After decades of living near one of America's most polluted waterways and advocating for remediation, locals were overjoyed that the Environmental Protection Agency had listed Newtown Creek as a Superfund site and would soon begin its massive cleanup operation. The designation came after years of grassroots activism by residents, who had meticulously documented the creek's contamination and made multiple requests for action.

Over the next decade, the transformation along Newtown Creek was striking. New waterfront parks and green spaces were designed to replace industrial wasteland (Kensinger, 2019). Luxury condominiums rose where oil refineries once stood. The air, thick for generations with petrochemical fumes, grew cleaner. But for many of the families who had endured decades of pollution, this environmental victory came with a bitter twist. As documented by the Furman Center, prices of residential properties in Greenpoint in 2023 increased by about 144% since 2009, while real median gross rent increased from $1,200 in 2006 to $2,330 in 2022, representing a 94.2% increase over the same period (NYU Furman Center, 2024). This stark increase in costs forced out some of the longtime residents who had fought for the creek's restoration, rendering environmental improvement a driver of displacement. Cohn (2014) termed this process "environmental gentrification" where the politics of sustainability and the market forces of gentrification overlap and reinforce each other to disempower low-income and marginalized communities.

At this point in time, we must pause to examine a critical cautionary tale from developed nations — one that developing countries would do well to heed as they pursue their own environmental transformations. The story of green gentrification represents perhaps the most sobering lesson about how even well-intentioned environmental improvements can perpetuate and deepen social inequalities. This phenomenon has its roots in the environmental justice movement of the 1980s when activists first began documenting the disproportionate burden of environmental hazards on low-income communities and communities of color. The irony is that the very success of these environmental justice campaigns sometimes leads to the displacement of the communities they were meant to protect.

The pattern is devastatingly consistent across Western cities. When industrial areas are cleaned up, when air quality improves, and when green spaces replace brownfields, property values inevitably rise. This seemingly positive transformation often triggers a displacement cascade that pushes out the very communities that suffered decades of pollution. In Chicago's Pilsen neighborhood, for instance, the remediation of lead-contaminated soil and the creation of new parks led to a significant increase in property values, displacing many Latino families who had lived there for generations (Bentley *et al.*, 2023; Hawthorne, 2019). Green gentrification demonstrates how environmental solutions, if not carefully managed, can exacerbate existing social disparities. Consider Portland, Oregon's Pearl District, once a polluted industrial zone transformed into an eco-friendly neighborhood showcase. While the area's green spaces increased and air quality improved dramatically, the original residents who had endured years of industrial pollution could no longer afford to stay and benefit from these improvements (Goodling *et al.*, 2015). Donovan *et al.* (2021) found that each one-percentage-point increase in tree canopy cover in Portland was associated with an $882 increase in the median sales price of single-family homes. In addition, neighborhoods with higher proportions of racial minorities and lower-income residents were less likely to benefit from tree-planting initiatives, highlighting the intersection of environmental improvement and social inequality. Similar patterns have played out in Barcelona's Poblenou district, once known as "the Catalan Manchester" because of its large concentration of factories, where industrial heritage has been replaced by gleaming eco-friendly developments as part of the 22@ project. Casamitjana (2023) detailed the following of Poblenou:

> Real estate speculation and the influx of highly skilled individuals with higher purchasing power have suffocated the lower-middle-class population, forcing them to move to other districts in the city with lower qualities of life: more concrete, more roads, less greenery and, consequently, poorer health.

Casamitjana (2023) points out that in most cities that go through such transitions, there are few public policies to prevent real estate speculation and the consequent exodus of old residents who can no longer keep up with the rapid rise in costs.

These Western examples offer crucial lessons for developing nations as they contemplate their own environmental cleanup efforts. Urban renewal projects must balance environmental improvement with social equity. India's Smart Cities Mission, for instance, must carefully consider these lessons as it pursues urban regeneration projects. The mission's focus on technological solutions and environmental improvements, while laudable, risks repeating the Western pattern of displacement if not properly managed. The experience of Seoul's Cheonggyecheon Stream restoration project, originally announced in 2002 and completed in 2005, offers an instructive case study. While the project successfully transformed a polluted urban waterway in Korea into a celebrated green space, it also led to significant displacement of small businesses and those who could not afford the sharply higher housing prices (Kim and Jung, 2019). This example particularly resonates with developing nations as they consider similar urban renewal projects.

The implications of green gentrification extend beyond housing prices to affect community cohesion and cultural identity. In Seattle's International District, the plans to create new housing projects, transit stations, and parks have contributed to the potential erosion of the neighborhood's historic Asian American character. Local businesses that served the community for decades are at risk of being replaced by upscale establishments catering to new residents, demonstrating how environmental improvements can inadvertently accelerate cultural displacement. Longtime journalist Ron Chew commented as follows (Pae, 2023):

> I think there's a danger in how people view the Chinatown-International District in that you know, there's a feeling that this is sort of a tourist destination where you can pick up some souvenirs and pick up your favorite Chinese exotic food. It's not simply restaurants. It's a neighborhood with people living here who have lived here for a long, long time and call this their home.

As we observe in Seattle's Chinatown-International District, which was named one of the "11 most endangered historic places" by the United States National Trust for Historic Preservation (2023), this pattern has been particularly pronounced in historically immigrant neighborhoods, where environmental improvements often coincide with the loss of ethnic businesses, community centers, and cultural institutions.

However, some cities have begun developing models for more equitable environmental improvement. In Barcelona's more recent initiatives, the city has implemented the Right to Housing Plan 2016–2025 that ensures the social function of housing and protects long-term residents from displacement. Similarly, Vancouver's Coalition of Progressive Electors has advocated for rent control measures to reduce the economic incentive for landlords to evict long-time tenants (COPE, n.d.). On June 11, 2024, the City Council of Vancouver approved a Vacancy Control Policy that restricted rent increases in single-room accommodation-designated buildings between tenancies (City of Vancouver, 2024). The popularity of Community Land Trusts, which are non-profit organizations that help provide affordable housing for low-income households in specific communities, also aided in empowering local communities in Canada in dealing with rising prices due to sustainability efforts (Bunce and Aslam, 2016). The *Parkdale Neighborhood Land Trust* in Toronto has successfully preserved affordable housing in an area experiencing rapid environmental improvement and development pressure.

The role of community organizing and grassroots activism has proven crucial in developing more equitable approaches to environmental improvement. In Oakland, California, the Asian Pacific Environmental

Network (APEN) has successfully advocated for environmental improvements while protecting affordable housing through community benefit agreements with developers. These agreements ensure that environmental cleanup projects include provisions for affordable housing, local hiring, and community services. In 2024, in a joint project with the City of Oakland, APEN received $225,000 to create a "Rooted and Resilient" Oakland Chinatown Plan. It is intended to facilitate the advancement of local priorities and the promotion of community resilience strategies.

These lessons become particularly crucial as developing nations deploy AI and automation technologies to address pollution issues. Green gentrification demonstrates how environmental improvements can deepen inequalities within nations. The challenge lies in ensuring that environmental benefits are distributed equitably across society. For developing nations, several key lessons emerge. First, environmental improvement projects must incorporate social protection measures from the outset, not as afterthoughts. This might include rent stabilization policies, community land trusts, or requirements for affordable housing retention in areas slated for environmental improvement. Second, community engagement must go beyond superficial consultation to include genuine power-sharing in decision-making processes. The failures of Western urban renewal projects often stemmed from top-down approaches that ignored local knowledge and needs. Third, the timing and phasing of environmental improvements matter. Rather than pursuing dramatic transformations that can trigger sudden property value spikes, gradual improvements coupled with social protection measures may allow communities to adapt while maintaining their residence. Fourth, environmental justice must be measured not just by physical improvement but by who benefits from these changes. The metrics of success should include not only environmental indicators but also measures of displacement and community stability. AI-powered predictive analytics can help identify areas at risk of displacement before environmental improvements begin, allowing for preemptive policy interventions. Machine learning algorithms can analyze patterns of displacement in previously gentrified areas to develop more effective anti-displacement strategies. That said, these technologies must be deployed with careful attention to equity and transparency to avoid perpetuating existing biases.

The implementation of AI-driven solutions has already shown promise in several pilot programs. For example, Seattle's "Equitable Development Monitoring Program" uses analytics to track displacement risk, combining data from various sources on housing cost burden, the affordability and availability of rental housing, foreclosures, tenant relocation cases, etc. (City of Seattle, n.d.). The program also monitors indicators like access to parks and open spaces, exposure to pollution, and access to public transit with night and weekend service that record not just information on how residents are doing on income, employment, or education but also other aspects of places affecting the quality of life and access to opportunity. Such monitoring can enable the identification of priority areas and the implementation of necessary intervention efforts to prevent the displacement of vulnerable residents, businesses, and community organizations.

As developing nations increasingly adopt technological solutions to address air pollution, these lessons become ever more crucial. The promise of AI-driven environmental monitoring and smart city technologies must be balanced against the risk of creating "green" spaces that serve only the affluent while pushing vulnerable populations into new pollution hotspots. The path forward requires a delicate balance between environmental improvement and social equity. Contemporary examples demonstrate the possibility of success when community engagement and equity are prioritized. For instance, Boston's Dudley Street Neighborhood Initiative has become a model for community-controlled development, successfully combining environmental cleanup with affordable housing preservation. At one point in history, this neighborhood had been afflicted with disinvestment, redlining, arson fires, and dumping (Nagel, 1990). Its innovative community-led efforts, like the 1986 "Don't Dump on Us!" campaign to end the illegal dumping of toxic wastes on vacant lots, have allowed the community to maintain control over development patterns while improving environmental conditions.

As developing nations stand at the crossroads of environmental transformation, they have a unique opportunity to learn from the West's mistakes and chart a more equitable course. The challenge lies not just in cleaning the air but in ensuring that everyone has the right to breathe it. Success in this endeavor will require a rethinking of how we measure environmental progress, moving beyond the simple metrics of physical improvement to include measures of social equity, community stability, and preservation of cultural heritage. This might include developing new indices that combine environmental quality metrics with displacement risk factors and cultural preservation indicators, providing a more holistic view of sustainable urban development.

References

Bently, C., Mirabile, H., Johansson, S., Rawal, P., Choo, T., and Ho, A. (2023). Gentrification in Pilsen and its impacts. *Podcast: The University of Chicago.* Retrieved on July 23, 2024 from https://mappingglobalchicago.rcc.uchicago.edu/2023-elpaseo/gentrification/.

Bunce, S. and Aslam, F. C. (2016). Land trusts and the protection and stewardship of land in Canada: Exploring non-governmental land trust practices and the role of urban community land trusts. *Canadian Journal of Urban Research, 25*(2), 23–34.

Casamitjana, L. (2023, August 14). "Green gentrification" — When environmental progress pushes the poor out of cities. *Worldcrunch.* Retrieved on February 10, 2024 from https://worldcrunch.com/green/green-gentrification-barcelona.

City of Seattle. (n.d.). Displacement risk indicators. *Equitable Development Monitoring Program.* Retrieved on February 10, 2024 from https://population-and-demographics-seattlecitygis.hub.arcgis.com/pages/displacement-risk.

City of Vancouver. (2024). *SRA Vacancy Control Policy.* Retrieved on October 12, 2024 from https://vancouver.ca/people-programs/rental-and-renter-protection.aspx.

Cohn, L. D. (2014). Strategic alliances in the battle against environmental gentrification: A case study on Greenpoint, NY. *MIT Department of Urban Studies and Planning.* Retrieved on February 10, 2024 from https://web.mit.edu/nature/projects_14/pdfs/2014-EnvironmentalGentrification-Cohn.pdf.

COPE. (n.d.). Rent control for Vancouver. *Coalition of Progressive Electors.* Retrieved on October 12, 2024 from https://www.copevancouver.ca/renters_platform.

Donovan, G. H., Prestemon, J. P., Butry, D. T., Kaminski, A. R., and Monleon, V. J. (2021). The politics of urban trees: Tree planting is associated with gentrification in Portland, Oregon. *Forest Policy and Economics, 124,* 102387.

Goodling, E., Green, J., and McClintock, N. (2015). Uneven development of the sustainable city: Shifting capital in Portland, Oregon. *Urban Geography, 36*(4), 504–527.

Hawthorne, M. (2019, May 25). Pilsen residents kept waiting for cleanup of toxic lead contamination. *Chicago Tribune.* Retrieved on February 10, 2024 from https://www.chicagotribune.com.

Kensinger, N. (2019, August 22). In Greenpoint, new waterfront parks will transform the Newtown Creek. *Curbed New York.* Retrieved on February 10, 2024 from https://ny.curbed.com/2019/8/22/20828195/greenpoint-brooklyn-newtown-creek-nyc-park-photo.

Kherani, F. (2010, November 17). Cuomo announces settlement for Greenpoint oil spill. *The Brooklyn Ink.* Retrieved on February 10, 2024 from http://brooklynink.org/2010/11/17/19537-cuomo-announces-settlement-in-greenpoint-oil-spill/.

Kim, H. and Jung, Y. (2019). Is Cheonggyecheon sustainable? A systematic literature review of a stream restoration in Seoul, South Korea. *Sustainable Cities and Society, 45,* 59–69.

Nagel, A. I. (1990). The Dudley street neighborhood initiative: A case study in community-controlled planning. *Doctoral dissertation, Massachusetts Institute of Technology.* Retrieved on February 10, 2024 from https://dspace.mit.edu/bitstream/handle/1721.1/69266/23535411-MIT.pdf?sequence=2.

National Trust for Historic Preservation. (2023, May 9). Discover America's 11 most endangered historic places for 2023. Retrieved on February 10, 2024 from https://savingplaces.org/stories/11-most-endangered-historic-places-2023.

NYU Furman Center. (2024, May 21). Greenpoint/Williamsburg BK 01. *Furman Center for Real Estate and Urban Policy.* Retrieved on July 23, 2024 from https://furmancenter.org/neighborhoods/view/greenpoint-williamsburg.

Pae, C. (2023, May 10). Residents of Seattle's Chinatown-International District fighting to preserve the neighborhood's culture. *King 5.* Retrieved on February 10, 2024 from https://www.king5.com/article/news/community/facing-race/seattle-chinatown-international-district-fights-preserve-neighborhood-culture/281-70e133df-87fc-4d87-9e6f-8efbeda1a339.

Chapter 10

Space Race for Clean Air: How Satellite Cities Reshape Atmospheric Politics

On August 26, 2019, Indonesian President Joko Widodo made a stunning announcement: Jakarta, a megacity of 10 million people choking on some of the world's worst air pollution, would no longer serve as Indonesia's capital (Lyons, 2019). Instead, the nation would build an entirely new capital city in East Kalimantan, on the island of Borneo. Estimates indicated that the move would cost about $32.7 billion, out of which the government would fund about 19% (Lyons, 2019). Named "Nusantara," an olden term for "outer islands," this planned city promised to be everything Jakarta wasn't — green, clean, and sustainable. While sinking ground levels in Jakarta were cited as the primary reason (the subsidence is occurring at an estimated 25 cm a year), the promise of an escape from the current capital's notorious air pollution played a crucial role in garnering public support.

This bold move reflects a growing trend: Rather than addressing pollution challenges in existing cities, nations are increasingly choosing to build entirely new urban centers from scratch. This approach to urban development raises critical questions about environmental justice, resource allocation, and the future of sustainable urban planning. To understand these implications, we must examine several major planned city projects worldwide and their varying approaches to environmental challenges.

South Korea's Songdo International Business District represents one of the earliest contemporary attempts at building a clean city. Songdo, being built from 2003–2025 on reclaimed land near Incheon, was designed with 40% green public space, extensive bicycle infrastructure, and a ban on traditional vehicles in certain areas. The city's strategic coastal location allows sea breezes to naturally ventilate the urban area, preventing the accumulation of air pollutants that plague nearby Seoul. Built at a cost of over US$40 billion, Overstreet (2021) notes that Songdo was originally conceptualized as a "completely sustainable, high-tech city that would plan for a future without cars, without pollution, and without overcrowded spaces. It was essentially a utopia that offered everything that Seoul didn't."

Songdo's comprehensive approach to green infrastructure offers valuable insights into integrated air quality management. The city's parks aren't merely recreational spaces — they form a sophisticated urban lung. A network of green corridors creates natural wind tunnels, while strategically placed vertical gardens act as biological air filters. The city's central park, measuring an impressive 100 acres, serves as a massive carbon sink. Even the building designs incorporate living walls and rooftop gardens, creating a 3D approach to urban air filtration.

However, Songdo's experience also reveals the challenges inherent in building satellite cities from scratch. Despite being one of the most expensive real estate development projects in contemporary times, Landis (2022) notes that Songdo is a "failed" project and it did not attract as many businesses and residents as it had initially estimated. Songdo's high-tech amenities, while making life more convenient, also unwittingly promote feelings of isolation and loneliness that prevent the fostering of a strong sense of community — a key ingredient in vibrant cities (Poon, 2018).

Egypt's New Administrative Capital provides another significant example of this trend. Located east of Cairo and estimated to cost around $58 billion, the city capitalizes on geographic advantages to escape the Nile Valley's notorious temperature inversions that trap polluted air. Its designers have incorporated lessons learned from Cairo's struggles, creating wide boulevards that act as wind corridors and positioning industrial zones downwind of residential areas. The city's elevation and distance from the Nile Delta's humidity create more favorable conditions for pollutant dispersal. The smart sustainable green city promises to offer a much higher quality of life compared to the polluted and congested Cairo, where population density is up to 50,000 per square mile (Lewis, 2024).

Perhaps the most ambitious of these green city projects is Saudi Arabia's Neom, a planned $500 billion city that promises to be entirely powered by renewable energy. With the development of "The Line," a linear city designed to have no cars, no streets, and zero carbon emissions, Neom represents the ultimate vision of engineered escape from air pollution. The project's massive scale and environmental promises have captured global attention, though construction has only recently begun. Even so, building a utopian city in the middle of a desert is no easy task, and the project's execution has raised serious humanitarian concerns. Construction workers reportedly had to endure 16-hour shifts for 14 days straight, sometimes with an unpaid, three-hour bus commute to get to the desert site (Akkad, 2024). More troublingly, a 2024 ITV documentary titled *Kingdom Uncovered: Inside Saudi Arabia* alleged that over 21,000 foreign workers have died in Saudi Arabia since 2017 during the construction of megaprojects like Neom that are a part of Mohammed bin Salman's Saudi Vision 2030.

The pursuit of these pristine urban environments comes at a staggering cost, both financial and social. When Indonesia announced Nusantara's $32 billion price tag, environmental justice advocates quickly pointed out that this sum could revolutionize Jakarta's public transportation system or fund comprehensive industrial emission controls. Instead, the money will build a clean-air haven for government officials and wealthy individuals while millions continue to breathe Jakarta's toxic air. Amir (2023) termed Nusantara a product of "techno-nationalist urbanism," rather than a decision embodying rational cost–benefit analysis. He suggested that in the Indonesian case, the technological and nationalist components are heavily infused with political forces, where the symbolic glory of bringing to reality a utopian city overrides the fundamental interests of its citizens. Even building the new toll road to Nusantara involves clearing out biodiverse forest areas, with adverse impacts on the local ecosystem (Gokkon, 2023). A petition on Change.org initiated in 2022 by the Narasi Institute and backed by 45 public figures titled "Mr. President, 2022–2024 is not the time to move the capital" has since garnered over 37,500 signatures. The petition called on the president to stop the plan to build Nusantara given that people were already in a difficult economic situation due to the COVID-19 pandemic. Relatedly, Ng (2022) pointed out that the government's financial situation was further complicated by the fact that it already "owes about US$17 billion in hidden debt to China through loans issued to its state-owned enterprises."

The social impacts extend beyond financial considerations. In East Kalimantan, ethnic communities have faced displacement without adequate compensation or respectful communication (Raharja, 2024). Although the Indigenous residents living on the lands in East Kalimantan inherited their plantations from their ancestors, they may not have written legal documents of land ownership. Now, they are being pushed out of their lands in the name of economic development and being disenfranchised. The leader of the Balik people, Sibukdin, noted the following in an interview with Al Jazeera (Washington and Hasibuan, 2023):

> We don't want to be relocated from the land of our ancestors. And we feel our land will be taken by the government. They said this capital is for the welfare of all Indonesians? But which Indonesians? We don't feel it's for us … They can easily erase our rights. Such is the greatness of people in authority. We consider our historical sites to be the source of our power. But they even moved the graves of our ancestors. The new capital is haunting us, and haunting the future of our children too.

However, despite protests, the building plans have continued. By June 2024, Phase I of the Nusantara project, which involved constructing core government buildings and civil servant housing, as well as groundbreaking for schools, universities, and sustainable energy centers, was about 80% complete (Lau, 2024).

Similar issues plague the Neom project. In the process of building "sustainable" Neom, many Indigenous residents have been forcibly displaced. Thomas and El Gibaly (2024) reported that more than 6,000 people have been moved and several villages in the Tabuk province of northwestern Saudi Arabia — mostly populated by the Huwaitat tribe — have been wiped off the map. Relatedly, a government order issued in April 2020 stated that the Huwaitat tribe included rebels and that continued resistance to eviction would be punishable by death, thereby licensing the use of lethal force against anyone who remained in their homes (Thomas and El Gibaly, 2024). As of 2023, a report found that "at least 15 members of the al-Huwaitat tribe have already been sentenced to prison terms of between 15 and an extraordinary 50 years, and at least five have been sentenced to death" (ALQST for Human Rights, 2023, p. 11). Alia Hayel Aboutiyah, a member of the tribe living in London, observed as follows (Michaelson, 2020):

> For the Huwaitat tribe, Neom is being built on our blood, on our bones. It's definitely not for the people already living there! It's for tourists, people with money. But not for the original people living there.

This pattern of environmental escapism to smart, green satellite cities reveals a troubling trend in urban development. In Egypt, the New Administrative Capital's gleaming towers and clean air will serve government workers and affluent residents, while Cairo's working class continues to endure some of the world's worst air quality. The project's budget could have funded extensive public transportation improvements or industrial retrofitting in Cairo, benefiting millions instead of a privileged few. These developments risk creating a new form of environmental apartheid — cities divided not by walls but by air quality. The wealthy retreat to manufactured clean-air oases while the poor remain trapped in increasingly polluted urban cores. This segregation has profound implications for public health, economic opportunity, and social mobility.

However, advocates of new sustainable smart cities have highlighted how the architectural and environmental engineering behind these planned developments embodies a sophisticated understanding of how urban design influences air quality. Each major project approaches the challenge of air pollution through multiple, interconnected strategies that have important implications for building sustainable solutions to our environmental problems. Consider the "green" carbon-neutral capital Nusantara's location in Indonesia, carefully chosen after extensive environmental impact studies. The site's natural wind patterns promise to sweep pollutants away from the city center, while its elevation above sea level helps avoid the temperature inversions that trap pollution in many Asian capitals. This attention to aerodynamics extends to the planned layout of buildings, with computational fluid dynamics models informing the positioning of every major structure to optimize airflow. In addition, the city would be designed in a way that 80% of journeys can be made by public transport, biking, or walking. Moreover, urban planners also paid close attention to other planned cities like Brasília, Canberra, and Malaysia's administrative center of Putrajaya in order to avoid repeating past mistakes.

As such, the debate over these new cities extends beyond their immediate impact on environmental justice. Supporters position them as vital laboratories for sustainable urban innovation, pointing to breakthrough technologies emerging from these ambitious projects. Masdar City's 45-meter-tall wind tower in Abu Dhabi stands as a testament to this potential — a modern reimagining of traditional Arabian *barjeel* that naturally cools and filters the air. The tower pulls in air from above, while sophisticated filters remove desert sand and pollutants, demonstrating how ancient wisdom can be enhanced by modern technology to address contemporary challenges (Hassan *et al.*, 2016). The LED lights that run up each of the three legs of the wind tower signal whether the city is doing well on its daily goal for energy consumption (green light) or whether the city has exceeded the daily limit by some amount (red light).

Yet this laboratory argument contains a troubling subtext — an implicit admission that our existing cities are beyond redemption. When South Korea poured $40 billion into Songdo's pristine streets, it tacitly suggested that Seoul's air quality problems were insurmountable. This defeatist approach threatens to create a dangerous precedent where wealthy nations abandon their polluted urban cores rather than invest in their renovation. The resources devoted to these new cities might have funded groundbreaking pollution control technologies or innovative public transportation solutions in existing metropolitan areas, benefiting millions more people.

The rise of clean-air havens signals a potentially dangerous shift in urban environmental policy. As wealthy nations increasingly opt to build anew rather than reform, they create a template that developing nations cannot afford to follow. The median African or South Asian city, already struggling with limited resources, cannot simply start fresh elsewhere. This disparity threatens to widen the global environmental justice gap, creating an urban future sharply divided between pristine planned cities and struggling legacy metropolises.

The environmental paradox of these projects adds another layer of complexity. Nusantara's construction has already raised alarming questions about deforestation in Borneo's sensitive ecosystems. Neom's massive scale threatens local villages and their limited water resources, while its construction generates significant carbon emissions. A cruel irony emerges: In our quest to build sustainable cities, we risk causing irreparable damage to the very environment we claim to value.

Brasília's story offers crucial insights into today's planned cities. When Brazil inaugurated its new capital in 1960, it represented the pinnacle of modernist urban planning — a city designed from scratch to escape the problems plaguing Rio de Janeiro, including its deteriorating air quality. Yet 60 years later, Brasília tells a more nuanced tale. While it initially achieved its environmental goals, its rigid zoning and car-dependent design eventually created new environmental challenges. Meanwhile, Rio's problems persisted, with the transfer of capital reducing the political urgency to address the old city's challenges. The parallel with today's clean-air havens is striking. Brasília's experience suggests that building new cities may temporarily avoid environmental problems but rarely solves them. Moreover, the departure of political and economic elites from troubled cities often reduces pressure for reform, leaving behind populations with diminished political capital to advocate for environmental improvements.

The tension between innovation and displacement extends beyond Indigenous communities to affect broader patterns of urban development. These projects often create situations where new, high-tech systems deliberately avoid integration with existing urban networks. In Nusantara's case, the planned smart grid and water management systems will operate independently from Jakarta's infrastructure, potentially drawing resources and expertise away from solving the capital's chronic problems. The technological achievements of new cities may come at the cost of widening the development gap between privileged enclaves and existing urban areas.

As we witness this new wave of urban development, we must confront fundamental questions about environmental justice and urban sustainability. The technology and innovation emerging from these new cities offer valuable lessons, but their implementation must extend beyond the boundaries of wealthy enclaves. Rather than creating islands of environmental privilege, we should focus on developing scalable solutions that can be retrofitted into existing urban centers. The path forward likely lies in synthesizing the insights from these new developments with the needs of existing cities. Masdar's wind towers could inspire natural ventilation solutions for older neighborhoods. Songdo's green corridor system could inform the redesign of existing urban parks. Neom's zero-emission transportation innovations could guide the transformation of current transit systems. Ultimately, the measure of these projects' success should not be how effectively they create escape hatches for the privileged, but how well their innovations can be democratized to benefit all urban residents. Clean air cannot remain a luxury good, accessible only to those who can afford to live in newly built environments. The true challenge lies not in building perfect cities from scratch, but in transforming our existing urban centers into healthier, more sustainable communities for all their residents.

References

Akkad, D. (2024, October 24). Saudi Arabia: Neom workers speak of '16-hour work days' in ITV undercover film. *Middle East Eye*. Retrieved on November 7, 2024 from https://www.middleeasteye.net/news/neom-line-workers-long-hours-accidents-anxiety.

ALQST for Human Rights. (2023, February). The dark side of Neom: Expropriation, expulsion and prosecution of the region's inhabitants. Retrieved on January 6, 2024 from https://alqst.org/en/post/the-dark-side-of-neom-expropriation-expulsion-and-prosecution.

Amir, S. (2023). Scrutinizing Nusantara: The making of an authoritarian city. *Southeast Asia Working Paper Series No. 5*. London School of Economics and Political Science.

Gokkon, B. (2023, April 10). To build its 'green' capital city, Indonesia runs a road through a biodiverse forest. *Mongabay*. Retrieved on January 6, 2024 from https://news.mongabay.com/2023/04/to-build-its-green-capital-city-indonesia-runs-a-road-through-a-biodiverse-forest/.

Hassan, A. M., Lee, H., and Yoo, U. (2016). From medieval Cairo to modern Masdar City: Lessons learned through a comparative study. *Architectural Science Review, 59*(1), 39–52.

Landis, J. D. (2022). A Case of Hubris-Songdo International Business District. In *Megaprojects for Megacities*, Northampton, MA: Edward Elgar Publishing, pp. 429–453.

Lau, J.M. (2024). The long road to Nusantara — shifting Indonesia's capital. *Lee Kuan Yew School of Public Policy, National University of Singapore*.

Lewis, N. (2024, March 20). A new city is rising in Egypt. But is it what the country needs? *Cable News Network (CNN)*. Retrieved on November 7, 2024 from https://www.cnn.com/world/egypt-new-administrative-capital-spc-intl/index.html.

Lyons, K. (2019, August 27). Why is Indonesia moving its capital city? Everything you need to know. *The Guardian*. Retrieved on January 6, 2024 from https://www.theguardian.com/world/2019/aug/27/why-is-indonesia-moving-its-capital-city-everything-you-need-to-know.

Michaelson, R. (2020, May 4). 'It's being built on our blood': The true cost of Saudi Arabia's $500bn megacity. *The Guardian*. Retrieved on January 6, 2024 from https://www.theguardian.com/global-development/2020/may/04/its-being-built-on-our-blood-the-true-cost-of-saudi-arabia-5bn-mega-city-neom.

Ng, J. (2022, March 18). Nusantara capital plans to carve out Jokowi's legacy. *East Asia Forum*. Retrieved on January 6, 2024 from https://eastasiaforum.org/2022/03/18/nusantara-capital-plans-to-carve-out-jokowis-legacy.

Overstreet, K. (2021, June 11). Building a City from Scratch: The Story of Songdo, Korea. *Arch Daily*. Retrieved on January 6, 2024 from https://www.archdaily.com/962924/building-a-city-from-scratch-the-story-of-songdo-korea.

Poon, L. (2018, June 22). Sleepy in Songdo, Korea's smartest city. *Bloomberg*. Retrieved on January 6, 2024 from https://www.bloomberg.com/news/articles/2018-06-22/songdo-south-korea-s-smartest-city-is-lonely.

Raharja, D.P. (2024, September, 19). The inaugural Independence Day ceremony in Nusantara: A milestone or a setback? *Heinrich Böll Stiftung: Southeast Asia*. Retrieved on November 17, 2024 from https://th.boell.org/en/2024/09/19/nusantara-milestone-setback.

Thomas, M. and El Gibaly, L. (2024, May 8). Neom: Saudi forces 'told to kill' to clear land for eco-city. *BBC Verify and BBC Eye Investigations*. Retrieved on November 7, 2024 from https://www.bbc.com/news/world-middle-east-68945445.

Washington, J. and Hasibuan, S. (2023, March 15). 'Like we don't exist': Indigenous fear Indonesia new capital plan. *Al Jazeera News*. Retrieved on January 6, 2024 from https://www.aljazeera.com/news/2023/3/15/like-we-dont-exist-indigenous-fear-indonesia-new-capital-plan.

Stewardship: Charting a Path Toward a Global Ethical Skyline

Chapter 11

For Nurrundere: Integrating Indigenous Wisdom into Sustainability

Decades ago, when European colonizers first arrived in Australia, they encountered a landscape shaped by sophisticated fire management techniques developed over thousands of years. The Indigenous practice of "fire-stick farming" involved carefully planned, low-intensity burns that prevented catastrophic bushfires and promoted biodiversity. However, colonial authorities, viewing these practices through the lens of European forestry science, dismissed them as primitive and destructive. The colonial perspective perceived Indigenous knowledge as superstitious gimmicks that needed to be replaced by "modern" Western approaches. This repudiation was particularly evident in the treatment of practitioners of cool and quick burning, who were often forcibly prevented from continuing their traditions, leading to a dangerous accumulation of fuel load in many areas. The consequences of this disruption became apparent in the devastating bushfires that would later plague the continent, particularly in regions where ancestral burning practices had been abandoned.

The indigenous practice of burning provides perhaps a compelling yet controversial example of ancient wisdom's relevance to contemporary air quality challenges. This sophisticated fire management system, developed over generations, demonstrates an intricate understanding of fire ecology and its impact on air quality. Given the specific context, by conducting controlled burns according to seasonal patterns and wind conditions, Indigenous communities effectively prevented catastrophic wildfires that would otherwise release massive amounts of pollutants into the atmosphere. This knowledge, when properly integrated with modern fire management protocols and applied in situations where appropriate, offers a powerful tool for reducing wildfire-related air pollution. Hoffman *et al.* (2021) suggested that Indigenous fire stewardship, though seemingly counterintuitive, can significantly decrease the severity of wildfires by reducing the amount of available fuels and increasing the fire resistance of vegetation. The Arnhem Land Fire Abatement project, for instance, has successfully combined traditional knowledge with modern carbon accounting methods, demonstrating how Indigenous practices can contribute to both environmental protection and economic development (Australian Government Clean Energy Regulator, 2024).

The colonial disregard for traditional knowledge extended to medicinal plants and biodiversity conservation. In many colonized regions, traditional practices of forest management that maintained medicinal plant populations were labeled as unscientific gathering. Colonial authorities imposed "scientific" forestry focused on timber production, leading to the loss of countless medicinal species and associated knowledge. Western scientific institutions often later "discovered" and patented compounds from these same plants, while discounting the sophisticated understanding that Indigenous peoples had developed about their characteristics. Furthermore, the "good intentions" of imposing Western scientific management masked economic exploitation. Colonial authorities used scientific rhetoric to justify the replacement of sustainable Indigenous practices with exploitative systems that prioritized short-term resource extraction over long-term sustainability. A striking example of this can be found in the story of the Pacific Yew tree, whose bark contains paclitaxel (Taxol), a compound used in cancer treatment. Indigenous peoples of the Pacific Northwest had long known of the tree's medicinal properties, but this knowledge was initially dismissed until Western researchers "discovered" its anti-cancer properties in the second half of the 20th century.

Similarly, the neem tree in India, traditionally used for various medicinal purposes including air purification, faced biopiracy attempts through patents filed by Western corporations, leading to lengthy legal battles to protect traditional knowledge rights (Bhattacharya, 2014). Gulyani and Singh (2010) detailed the process (p. 143):

In 1994 the EPO [European Patent Office] granted European Patent No. 0436257 to the US Corporation W.R. Grace and USDA for a method for controlling fungi on plants by the aid of hydrophobic extracted Neem oil.

In 1995, a group of international NGOs and representatives of Indian farmers filed a legal opposition against the patent. They submitted evidence that the fungicidal effect of extracts of neem seeds had been known and used for centuries in Indian agriculture to protect crops, and thus the invention (…) was not novel.

The patent was ultimately revoked in 2000, about six years after it was originally granted. This case exemplifies the profound injustice faced by Indigenous communities who must wage expensive legal battles to defend knowledge that their ancestors freely shared and preserved for centuries — knowledge that was intended not as intellectual property to be owned, but as a collective heritage to benefit all of humanity.

Several regions today face environmental challenges directly traceable to the disruption of traditional management systems during colonization. The irony lies in how colonial authorities misconstrued the concept of science itself. True scientific inquiry involves observation, testing, and adaptation — precisely the processes that Indigenous peoples had engaged in over generations to develop their practices. Their perspective encompassed a circular view of environmental management, recognizing the intricate interconnectedness of natural systems. Consider the Māori concept of *kaitiakitanga* (Te Ahukaramū, 2007), which embodies the principle that humans are not separate from nature but integral parts of it, bearing sacred responsibilities to maintain environmental balance. This worldview starkly contrasts with the linear, extractive approach that has dominated modern industrial societies and led to our current crisis. A utilitarian understanding of landscape with the aim of commodification is fundamentally different from a holistic attitude toward understanding nature that does not draw clear boundaries and pays attention to complex webs of ecological relationships. For example, the traditional Polynesian concept of "reading" the ocean through observation of wave patterns, cloud formations, and wildlife behavior represents a sophisticated form of environmental science that Western researchers are only beginning to appreciate.

As urban pollution levels rise, wildfires increase in frequency and intensity, and industrial emissions continue to grow, we face a future where the very air we breathe threatens our existence. The integration of Indigenous knowledge into modern air quality management strategies is not merely an academic exercise but a survival imperative. In California, for instance, recent collaborative efforts of the government with the Karuk and the Yurok Tribes to provide targeted understory fire-based forest treatments have resulted in significant associated benefits (Marks-Block *et al.*, 2019). The authors noted the following (Marks-Block *et al.*, 2019):

> Fire exclusion policies forced California Indian communities and forest managers to curtail their routine cultural and prescribed burning practices. Despite these policies, Karuk and Yurok basketweavers retained their knowledge, maintained their practices and, most importantly, developed several innovative techniques to replicate fire's effects on hazelnut to produce essential basketry materials.

These partnerships demonstrate how traditional ecological knowledge can be effectively integrated with modern scientific approaches to address contemporary environmental challenges.

Let's think about the profound wisdom embedded in the Haudenosaunee (Iroquois) agricultural practice known as the *Three Sisters*. This sophisticated system of planting corn, beans, and squash together demonstrates an intimate understanding of ecological relationships that naturally maintain air quality. The practice eliminates the need for chemical fertilizers, maximizes soil nitrogen naturally, and prevents soil erosion and dust — all contributing to cleaner air through sustainable agriculture. However, colonial authorities had historically dismissed the sophisticated polyculture systems developed by Indigenous peoples, replacing them with European-style monoculture. The colonizers viewed their approach as superior because it aligned with the European scientific understanding of the time, despite its unsuitability for local ecosystems. In reality, Three Sisters companion planting can sequester more carbon than conventional monoculture systems because of its mimicry of natural ecological relationships — the corn provides a climbing structure for the beans, which fix nitrogen in the soil, while the squash's broad leaves suppress weed growth and reduce water evaporation.

Indigenous frameworks and traditional knowledge of architecture and urban planning principles offer equally valuable insights for modern cities struggling with air pollution. The sophisticated ventilation systems found in Persian wind towers (*badgirs*) and the naturally ventilated courtyards of Pueblo settlements demonstrate an advanced understanding of air circulation that could revolutionize modern green building design. These time-tested solutions, developed through centuries of observation and adaptation, offer sustainable alternatives to energy-intensive modern air-conditioning systems. For instance, the wind towers of Iran's desert city of Yazd, which date back to the 14th century, have been providing natural cooling and ventilation for centuries, operating without electricity while maintaining comfortable temperatures and air quality (Abdolhamidi, 2018). Modern architects are increasingly studying these traditional designs to develop more sustainable urban environments. The principles behind these ancient systems are being incorporated into contemporary "smart city" designs in places like Masdar City, UAE, where traditional Arabic architectural elements are combined with modern technology to create energy-efficient, naturally ventilated spaces.

In South and Southeast Asia, traditional practices continue to offer valuable insights for modern environmental management. In India, for example, the ancient practice of using *neem* (Azadirachta indica) leaves for air purification has gained renewed interest. Neem leaves have been traditionally burned or hung in homes to purify the air and repel insects. Recent scientific studies, such as the work by Rajendra Prasad and Samendra Prasad (2018), have corroborated the air-purifying properties of *neem* due to the presence of a large number of volatile organic chemicals, showing *neem's* potential in absorbing particulate matter and certain gaseous pollutants. Madiraju *et al.* (2020) experimented with air purifiers designed with eco-friendly materials and adsorbents prepared from plant extracts including neem bark, mango bark, orange peel powder, and neem leaf powder. Their study on the reduction in indoor pollution levels by using such air purifiers in five separate locations showed positive results. This convergence of traditional wisdom and scientific validation opens up possibilities for developing low-cost, locally appropriate air purification solutions in urban areas of developing countries.

In East Asia, the traditional practice of using bamboo in construction and daily life offers insights into sustainable materials that can indirectly impact air quality. Bamboo, a fast-growing grass, has been used for millennia in countries like Indonesia, Vietnam, and the Philippines. Its rapid growth and high carbon sequestration capacity make it an excellent tool for combating air pollution. Song *et al.* (2011) drew attention to the importance of bamboo forests in sequestering carbon and mitigating pollution. Integrating this traditional material into modern green building practices and reforestation efforts could significantly contribute to air quality improvement in developing countries. The authors note that as a sustainable carbon sink, bamboo is receiving increasing attention in Chinese afforestation and reforestation projects (Song *et al.*, 2011). In 2023, China's National Development and Reform Commission and other three central government agencies launched the "Three-year action plan for promoting bamboo as an eco-friendly substitute for plastics" (Yake, 2023). The traditional knowledge of bamboo cultivation and management, passed down through generations in rural communities, is now being systematically documented and integrated into urban forestry programs across China.

In China, the ancient practice of Feng Shui, often dismissed as mere superstition, contains principles that align with the modern understanding of air circulation and quality in built environments. The emphasis on proper orientation of buildings, the flow of "qi" (which can be interpreted as air or energy), and the integration of natural elements in living spaces all contribute to better indoor air quality. While not a direct solution to outdoor air pollution, these principles can inform urban planning and architectural designs that promote better air circulation and reduce the concentration of indoor air pollutants. Research by Xu (2022) highlights how Feng Shui principles focus on promoting good *qi* by ensuring a residence has adequate sunlight, fresh air, and decent moisture. Han and Lin (2023) observed that in China, traditional quadrangular buildings were designed in a way that helped to maintain good indoor air quality, which is lacking in many crowded modern constructions. Modern studies of traditional Chinese courtyard homes have revealed sophisticated passive ventilation systems that maintain optimal airflow patterns throughout the year. These designs typically incorporate features such as graduated roof heights, strategic window placement, and internal courtyards that facilitate the flow of air. Contemporary architects working on sustainable building projects in Beijing and Shanghai have successfully adapted these principles to modern high-rise developments, resulting in significant reductions in energy consumption for ventilation while maintaining superior indoor air quality.

The integration of these traditional knowledge systems into modern air pollution management strategies, however, is not without challenges. One key issue lies in the epistemological divide and the difficulty of reconciling different worldviews. N. Scott Momaday (1968/1999), writing in his book *House Made of Dawn*, began the story by highlighting the connection between man and nature, "(...) The land was still and strong. It was beautiful all around. Abel was running … He was running, running. He could see the horses in the fields and the crooked line of the river below" (preface). This connection to the land represents precisely the kind of environmental knowledge that colonial authorities dismissed as primitive or unscientific. Abel's story was not just about personal displacement, but the broader consequences of forcing Indigenous people to abandon their traditional understanding of environmental relationships. In Momaday's book, the urban landscape, with its concrete and artificial boundaries, stood in stark contrast to the integrated simplicity represented by Abel's grandfather's world. The novel's portrayal of Abel's displacement mirrors the current challenge of translating traditional environmental knowledge into terms that modern institutions can understand and implement. Attempts to force Indigenous wisdom into existing scientific paradigms rather than allowing it to inform and reshape those paradigms could lead the former to lose its very essence. This disconnect is particularly evident in contemporary environmental impact assessments, which often fail to capture the holistic nature of Indigenous environmental knowledge. For instance, while Western scientific methods might measure air quality through specific pollutant concentrations, Indigenous knowledge systems might consider the broader ecological context, including the health of indicator species, seasonal patterns, and the interconnected effects on community well-being.

Another significant hurdle is the lack of documentation and scientific validation of many traditional practices. To address this, collaborative research initiatives involving Indigenous communities, environmental scientists, and policymakers are essential. Another challenge lies in adapting traditional practices to the scale and complexity of modern urban environments in developing countries. This requires a careful balance between preserving the essence of traditional wisdom and innovating to meet contemporary challenges. The concept of hybrid knowledge, which combines traditional and scientific knowledge, offers a promising framework for this integration. For example, in Australia, the government is implementing the Indigenous Rangers Program to acknowledge and support "First Nations peoples' unique, critical and continuing role in managing and protecting Australia's natural and cultural heritage" (NIAA, n.d.).

The application of traditional environmental knowledge to urban air quality management represents a promising frontier in sustainable city development. Traditional practices of creating urban green spaces, managing water bodies, and designing buildings can be adapted to modern city contexts to improve air quality. For instance, the traditional Japanese concept of "Shinrin-yoku" (forest bathing) has inspired the creation of urban forests that serve as natural air purifiers while providing psychological benefits to communities. "Shinrin-yoku" was formally defined by the Japanese Ministry of Agriculture, Forestry and Fisheries in 1982 as "the process of soaking up the sights, smells and sounds of a natural setting to promote physiological and psychological health" (Wisniewski, 2017). In this regard, Park *et al.* (2010) found that such forested environments helped "promote lower concentrations of cortisol, lower pulse rate, lower blood pressure, greater parasympathetic nerve activity, and lower sympathetic nerve activity" (p. 18) compared to typical city environments. Similarly, the Aztec practice of creating "chinampas" (floating gardens) has been adapted for modern urban agriculture projects that help reduce urban heat island effects and improve local air quality. However, Merlín-Uribe *et al.* (2013) deplored the contemporary trend of substituting chinampas with plastic greenhouses because of the related adverse impacts on the landscape, local environment, and traditional culture.

As we look to the future, the successful integration of traditional ecological knowledge into modern air quality management requires a multifaceted approach that addresses both institutional and practical challenges. The establishment of formal mechanisms for incorporating traditional knowledge into environmental policymaking stands as a crucial first step, ensuring that Indigenous voices are not merely consulted but are integral to decision-making processes. This needs to be coupled with robust collaborative research programs that bring together traditional knowledge holders and environmental scientists, creating a synergistic approach that bridges different ways of understanding and managing air quality. The preservation and documentation of traditional ecological knowledge, conducted with the utmost respect for Indigenous intellectual property rights and cultural protocols, should be prioritized before this invaluable wisdom is lost to time. Furthermore, the development of educational programs that seamlessly weave traditional ecological knowledge with modern environmental science could foster a new generation of environmental managers capable of working effectively across different knowledge systems. These theoretical foundations should be potentially complemented by practical applications through pilot projects that demonstrate the real-world value of traditional knowledge in modern contexts, particularly in urban environments where air quality challenges are most acute. The journey toward cleaner air and more sustainable environmental management needs to acknowledge and incorporate the wisdom of traditional ecological knowledge, representing not just a practical necessity but a moral imperative to respect and learn from cultures that have maintained sustainable relationships with their environments for millennia. Like the story of the supreme celestial being Nurrundere, whose rainbow bridge connects earth and sky (Pettazzoni, 1924), we must build bridges between traditional and modern knowledge systems, recognizing that both are essential parts of the same environmental story.

The sophisticated nature of Indigenous ecological knowledge is perhaps best exemplified in traditional foodways, which demonstrate how sustainable practices can be deeply embedded within cultural systems. The Navajo "Corn Pollen Path" illustrates how traditional food systems transcend mere agricultural practice to encompass spiritual wisdom and environmental stewardship. These teachings emphasize that corn isn't merely a food source — it's a living entity requiring relationship and reciprocity. Each stage of corn's growth cycle is accompanied by specific ceremonies and prayers, reflecting a deep understanding of natural cycles and the interconnectedness of all things in the natural world. This worldview stands in stark contrast to industrial agriculture's treatment of crops as commodified tools to make money, offering important lessons for reducing agricultural air pollution in an era of climate change and environmental degradation. The revival of traditional food systems through Indigenous-led food sovereignty movements could play a crucial role in improving regional air quality through ecosystem restoration.

By drawing on the wisdom accumulated over generations and combining it with modern scientific understanding, we can develop more holistic, culturally appropriate, and sustainable solutions to air quality challenges. This approach not only addresses the immediate problem of air pollution but also promotes the preservation of cultural heritage and biodiversity. In this integration lies hope for a future where clean air is not a privilege but a right, and where human activities exist in harmony with natural systems. Our survival depends on our willingness to learn from those who have maintained this harmony for generations.

References

Abdolhamidi, S. (2018, September 27). An ancient engineering feat that harnessed the wind. *BBC News.* Retrieved on August 22, 2024 from https://www.bbc.com/travel/article/20180926-an-ancient-engineering-feat-that-harnessed-the-wind.

Australian Government Clean Energy Regulator. (2024, March 23). Arnhem Land Fire Abatement. Retrieved on August 22, 2024 from https://cer.gov.au/news-and-media/case-studies/arnhem-land-fire-abatement.

Bhattacharya, S. (2014). Bioprospecting, biopiracy and food security in India: The emerging sides of neoliberalism. *International Letters of Social and Humanistic Sciences*, (12), 49–56.

Gulyani, M. and Singh, N. (2010). Bio-piracy: The appropriation of traditional knowledge. *International Journal of Business Economics and Management Research*, *1*(2), 137–149.

Hoffman, K. M., Davis, E. L., Wickham, S. B., Schang, K., Johnson, A., Larking, T., Lauriault, P., Le, N., Swerdfager, E., and Trant, A. J. (2021). Conservation of Earth's biodiversity is embedded in Indigenous fire steward-ship. *Proceedings of the National Academy of Sciences*, *118*(32), doi: https://doi.org/10.1073/pnas.2105073118.

Han, K. T. and Lin, J. K. (2023). Empirical and quantitative studies of Feng Shui: A systematic review. *Heliyon*, *9*(9), e19532.

Madiraju, S. V. H., Raghunadh, P. G., and Kumar, K. R. (2020). Prototype of eco-friendly indoor air purifier to reduce concentrations of CO_2, SO_2 and NO_2. *Nature Environment and Pollution Technology an International Quarterly Scientific Journal*, *19*(2), 747–753.

Marks-Block, T., Lake, F. K., and Curran, L. M. (2019). Effects of understory fire management treatments on California Hazelnut, an ecocultural resource of the Karuk and Yurok Indians in the Pacific Northwest. *Forest Ecology and Management*, *450*, 117517.

Merlín-Uribe, Y., González-Esquivel, C. E., Contreras-Hernández, A., Zambrano, L., Moreno-Casasola, P., and Astier, M. (2013). Environmental and socio-economic sustainability of chinampas (raised beds) in Xochimilco, Mexico City. *International Journal of Agricultural Sustainability*, *11*(3), 216–233.

Momaday, N. S. (1968/1999). *House made of dawn*. New York, NY: Perennial Classics.

NIAA. (n.d.). Indigenous Rangers. *National Indigenous Australian Agency*. Retrieved on August 22, 2024 from https://www.niaa.gov.au.

Park, B. J., Tsunetsugu, Y., Kasetani, T., Kagawa, T., and Miyazaki, Y. (2010). The physiological effects of Shinrin-yoku (taking in the forest atmosphere or forest bathing): Evidence from field experiments in 24 forests across Japan. *Environmental Health and Preventive Medicine*, *15*, 18–26.

Pettazzoni, R. (1924). The chain of arrows: The diffusion of a mythical motive. *Folklore*, *35*(2), 151–165.

Prasad, R. and Prasad, S. (2018). Neem and the environment. *International Journal of Plant and Environment*, *4*(01), 1–9.

Song, X., Zhou, G., Jiang, H., Yu, S., Fu, J., Li, W., Wang, W., Ma, Z., and Peng, C. (2011). Carbon sequestration by Chinese bamboo forests and their ecological benefits: Assessment of potential, problems, and future challenges. *Environmental Reviews*, *19*, 418–428.

Te Ahukaramū Charles Royal. (2007, September 24). Kaitiakitanga — guardianship and conservation. *Te Ara — The Encyclopedia of New Zealand.* Retrieved on August 22, 2024 from https://teara.govt.nz/en/kaitiakitanga-guardianship-and-conservation.

Wisniewski, A. (2017, June 22). Shirin-Yoku: Why forest bathing became a global health phenomenon. *American Forests.* Retrieved on August 22, 2024 from https://www.americanforests.org/article/shirin-yoku-why-forest-bathing-became-a-global-health-phenomenon/.

Xu, P. (2022). Healthy living in the built environment in light of Feng-shui. *International Journal of e-Healthcare Information Systems, 8*(1), 196–202.

Yake, L. (2023, November 30). China launches three-year plan for promoting bamboo as eco-friendly substitute for plastics. *Enviliance Asia.* Retrieved on August 22, 2024 from https://enviliance.com/regions/east-asia/cn/report_11208.

Chapter 12

Ground-Up Blue: Local Movements Against Gray Skies

Those who contemplate the beauty of the earth find reserves of strength that will endure as long as life lasts.

— Rachel Carson, *Silent Spring*

In July 1943, Los Angeles residents woke to what they initially thought was a Japanese chemical attack in the midst of the World War II. A thick blanket of smog had descended upon the city and the air "smelled like bleach" (Eschner, 2017), burning their eyes and throats, reducing visibility to three blocks, and forcing many to wear gas masks. This "gas attack from Japan" (McNully, 2010) turned out to be the city's first recognized episode of severe photochemical smog, created by the interaction of industrial emissions, vehicle exhaust, and sunlight. In response, city officials tried temporarily shutting down the "obvious culprit" — a nearby butadiene plant — but the smog still persisted. The plant's closure served as an early lesson in the complexity of urban air pollution — simple, single-source solutions would prove inadequate for addressing this multi-faceted problem. In 1947, the Los Angeles County Board of Supervisors established the first-of-its-kind Air Pollution Control District in the nation, which not only regulated oil refineries and power plants but also hired scientists to look into the sources of the smog, gather evidence, and provide a blueprint for action.

Eleven years after the 1943 incident, when the smog in Los Angeles had become unbearable for many, thousands of residents gathered together at the Pasadena Civic Auditorium and formed the Anti-Smog Action Committee, with the goal to "take the problem of smog out of the scientific laboratories and into the hands of the people" (*Distillations Podcast*, 2018). Their activism captured media attention and helped transform air quality from an abstract scientific concern into a visible public issue. This later petered out to give way to an organization called *Stamp Out Smog* (SOS). Started in 1958 by a group of concerned local women, "SOS soon became the most significant activist group battling smog" (Turner, 2019). During the 1960s, SOS used evocative public demonstrations and letter-writing campaigns to call for change (Turner, 2019). This early example of citizen mobilization against air pollution would later serve as a model for grassroots movements in developing countries, where communities faced similar — and often worse — air quality crises. The emotional landscape of these movements reveals a complex interplay of hope and frustration. Public health records and community surveys from various movements around the world document the profound psychological impact of living with severe air pollution — the anxiety of parents watching their children struggle to breathe, the frustration of communities facing institutional indifference, and the determination to create change despite seemingly overwhelming odds.

The emergence and evolution of local grassroots movements against air pollution in developing countries represent more than just a shift in environmental advocacy — they embody the indomitable human spirit in the face of seemingly insurmountable challenges. These movements often emerged in contexts where environmental protection was seen as a luxury that developing economies couldn't afford. Activists had to challenge not only specific polluters but also deeply entrenched narratives about the inevitable environmental costs of economic progress. Behind every protest, every petition, and every small victory are ordinary citizens who refused to accept the *status quo*, even when victory seemed impossible. Like the description of the protagonist in William Somerset Maugham's *The Moon and the Sixpence*, these activists' reputations do not shine because of external trappings or official positions, but because their goodness is deeply rooted, unshaken by the winds of opposition or the rain of setbacks.

The daily reality of these movements in developing countries reveals a profound intersection of necessity and moral conviction. Street vendors, factory workers, parents, and grandparents — people from all walks of life — have transformed their personal experiences with air pollution into collective action. These movements emerged not from professional activists or established organizations, but from communities facing the direct impact of contaminated air on their daily lives. These everyday workers adapt their activism to the rhythms of working life — holding meetings during off-hours, conducting health surveys on weekends, and using lunch breaks for community outreach. This integration of activism into daily routine demonstrates a profound understanding that lasting change requires sustained, strategic effort rather than sporadic gestures. The transformation of private suffering into public advocacy has become a defining characteristic of air pollution activism in developing nations.

The roots of such movements in developing countries can be traced back to the latter half of the 20th century, often intertwined with broader social and political movements. One of the earliest and most influential examples is the Chipko movement in India, which began in the 1970s. While primarily focused on forest conservation, the Chipko movement set a precedent for community-led environmental activism in India that would later influence air pollution-focused initiatives. The movement gained international attention when village women in the Himalayan region of Uttarakhand hugged trees to prevent felling, giving rise to the name "Chipko" (meaning "to stick" in Hindi). The movement's success in mobilizing local communities, particularly women, to protect their environmental resources demonstrated the power of grassroots action in effecting change (Gosai *et al.*, 2024). The prominent role of women in air pollution activism reflects a broader pattern in environmental justice movements. From the mothers who initiated Los Angeles' SOS to the women who led the Chipko movement, female activists have consistently been at the forefront of air quality advocacy. This leadership often emerges from women's roles as primary caregivers witnessing the health impacts of pollution on their families and communities.

In the context of air pollution, one of the most notable early grassroots movements emerged in Mexico City in the 1980s. Faced with severe air pollution that earned the city the moniker "most polluted city on the planet" (Air Quality Life Index, 2023), local citizens began to organize and demand action. The pollution levels were so extreme that birds would fall dead from the sky mid-flight. De Paola (2019) observes that the 1990s saw the United Nations label Mexico City "the most dangerous city in the world for children." The grassroots movement in the country gained momentum following a particularly severe air pollution episode in 1986. Citizen groups such as the *Movimiento Ecologista Mexicano* (Mexican

Ecological Movement) played a crucial role in pressuring the government to implement measures such as mandatory catalytic converters for vehicles and the relocation of heavy industries away from the city center. Their endeavors also helped in the adoption of the "Hoy No Circula" (No Drive Days) program in 1989, which restricted private vehicle usage based on license plate numbers (Guerra and Millard-Ball, 2017). These efforts contributed to significant improvements in Mexico City's air quality over the subsequent decades.

The moral courage underlying these movements becomes particularly evident in contexts where public protest carries significant risks. In countries with restrictive political environments, activism often takes more subtle forms — community education programs, health documentation efforts, and informal networks for sharing air quality information. These approaches demonstrate how determination and creativity can overcome institutional barriers. In China, where rapid industrialization has led to severe air pollution in many cities, grassroots movements have played a significant role in recent years despite the challenges posed by the political system. The turning point came in 2008 when the US Embassy in Beijing began publishing air quality data ahead of the Summer Olympics, revealing the true extent of the city's pollution problem to the international community (Roberts, 2015). Subsequently, in 2010, Roberts (2015) noted the following:

> Beijing's air quality was deemed "crazy bad" by the Embassy when the pollution exceeded the bounds of the EPA's air quality index. This inadvertently undiplomatic tweet reached a growing audience via third-party apps that circumvented China's twitter firewall (…) Beijing residents, dissatisfied with the crudeness of China's air quality monitoring efforts, put pressure on Chinese officials to acknowledge the scale of the problem and start taking proactive measures to tackle it.

This information catalyzed public awareness and led to the formation of numerous citizen-led initiatives. One notable example is the *Under the Dome* documentary by Chai Jing, a former China Central Television investigative journalist, released in 2015. Chai's narrative encompassed a string of institutional failures in China, including "incompetent government departments that are unable to execute environmental standards and regulation, and ambiguous environmental laws and policies that are ineffective in terms of tackling actual pollution cases" (Pan, 2017, p. 20). Within a few days of its release, the documentary was viewed 300 million times on Tencent. The film, which went viral on Chinese social media before being censored, sparked widespread public discussion about air pollution and put pressure on the government to take more decisive action (Deng and Peng, 2018). China's Environment Protection Minister Chen Jining acknowledged that *Under the Dome* played "an important role in promoting public awareness of environmental health issues" (Burki, 2015).

Activists like Chai Jing understand that their work may not bring immediate results; they may not live to see completely clear skies. Like the artist in Maugham's *The Moon and the Sixpence*, their worth isn't measured by immediate recognition or success, but by their unwavering commitment to a greater cause. This understanding is reflected in the long-term planning documents of various movements, which often span generations in their vision and scope. The challenges these movements face — limited resources, political opposition, and bureaucratic inertia — are not just obstacles to overcome but tests of collective moral courage. When authorities restrict information about air quality, communities respond by developing their own monitoring systems. When industrial interests push back against emissions regulations, local groups counter with increasingly sophisticated documentation of health impacts, transforming personal experiences into compelling data.

The rise of social media and digital technologies has provided new tools for grassroots movements to organize, share information, and exert pressure on authorities. In Thailand, the *Chiang Mai Breathe Council*, an independent body founded in 2019, brings together local thinkers and academics to put pressure on the government to fund clean air initiatives and make much-needed regulatory changes, as well as to initiate a social movement to raise awareness and educate the residents (Kemasingki, 2020). The group's online campaigns have been instrumental in raising awareness about the health impacts of air pollution and pushing for more stringent regulations on vehicle emissions and industrial pollutants.

In many developing countries, grassroots movements have increasingly focused on the intersection of air pollution with other social and economic issues. In South Africa, for instance, local communities in the Highveld region have organized to address the air pollution caused by coal-fired power plants. Organizations like the Highveld Environmental Justice Network have meticulously documented the health impacts on local communities (Centre for Environmental Rights *et al.*, 2017). These movements not only advocate for cleaner air but also highlight issues of environmental injustice, as the most severely affected communities are often low income and historically disadvantaged. The impact of these grassroots movements extends beyond their immediate local contexts. They have played a crucial role in shaping national and even international discourse on air pollution. The success of local initiatives in cities like Mexico City and Beijing has provided models for citizen engagement that have been adapted and replicated in other developing countries. Moreover, these movements have contributed to the growing global recognition of air pollution as a critical public health and environmental issue, influencing international agreements and cooperation on air quality management.

The evolution of these movements reflects a broader transformation in environmental activism — from isolated local struggles to interconnected global campaigns. While each movement remains rooted in local concerns and contexts, they increasingly share strategies, resources, and solidarity across borders. The *C40 Cities Climate Leadership Group*, a network of 96 cities making up over 22% of the global economy, has been particularly instrumental in facilitating this knowledge exchange. This global network of local actions has created a new model of environmental advocacy, one that combines the moral authority of grassroots activism with the power of international cooperation.

What emerges from our historical records is a narrative of persistent hope and incremental victory. Every improvement in air quality standards, every new regulation enacted, and every successful community initiative represents a triumph of collective will over institutional inertia. The legacy of these movements extends far beyond measurable improvements in air quality. They have demonstrated that organized communities, working together with extraordinary persistence, can challenge powerful interests and create lasting change. They have shown that grassroots activism isn't just about protest — it's about creating alternative visions of the future and working steadily toward them. The impact of these movements can be measured not just in reduced pollution levels, but in transformed social attitudes. In South Korea, the success of anti-pollution campaigns in Seoul led to the mainstreaming of environmental concerns in national politics. The *Korean Federation for Environmental Movement* (KFEM), originally founded in 1993 by eight nationwide groups, has grown to become one of Asia's largest civic environmental organizations, with 52 local chapters, five specialized institutions, and five cooperative institutions (KFEM, n.d.). Their evolution from focused air quality advocacy to broader environmental leadership illustrates how grassroots movements can reshape national priorities.

In the end, these movements tell us resoundingly that the most profound changes often begin not with grand gestures, but with collective acts of courage and tenacity. Their story is not one of uninterrupted victory, nor of noble defeat, but of something far more important — the daily choice to continue struggling for a better life, regardless of the odds. Like Indigenous communities who have maintained their traditional ecological knowledge through generations of adversity, these movements demonstrate the power of sustained collective action in preserving and protecting the fundamental right to clean air. In this way, they have indeed left a legacy that will stand the test of time — not in monuments or proclamations, but in the fundamental understanding that organized communities, working together with extraordinary persistence, can change the world.

References

Air Quality Life Index. (2023). Policy impact — Mexico City: ProAire (1990). *The University of Chicago*. Retrieved on January 11, 2024 from https://aqli. epic.uchicago.edu/policy-impacts/mexico-city-proaire-1990.

Burki, T. K. (2015). Smokestacks and censorship in China. *The Lancet Respiratory Medicine, 3*(6), 432.

Centre for Environmental Rights, groundWork, and Highveld Environmental Justice Network. (2017). Broken Promises: The failure of South Africa's priority areas for air pollution — time for action. Retrieved on January 11, 2024 from https://groundwork.org.za/wp-content/uploads/2022/07/broken-promises.pdf.

De Paola, S. (2019, February 14). A breath of fresh air for Mexico City. *Julius Baer*. Retrieved on January 11, 2024 from https://www.juliusbaer.com/en/insights/future-cities/a-breath-of-fresh-air-for-mexico-city/.

Deng, X. and Peng, S. (2018). Trust, norms and networks in social media environmental mobilization: A social capital analysis of Under the Dome in China. *Asian Journal of Communication, 28*(5), 526–540.

Distillations Podcast. (2018). Fighting smog in Los Angeles. *Science History Institute Museum & Library*. Retrieved on January 11, 2024 from https://www.sciencehistory.org/stories/distillations-pod/fighting-smog-in-los-angeles/.

Eschner, K. (2017, July 26). This 1943 "Hellish Cloud" was the most vivid warning of LA's smog problems to come. *Smithsonian Magazine*. Retrieved on

January 11, 2024 from https://www.smithsonianmag.com/smart-news/1943-hellish-cloud-was-most-vivid-warning-las-smog-problems-come-180964119/.

Gosai, H. G., Sharma, A., and Mankodi, P. (2024). Major Environmental Activism in India: Past and Present. In *Environmental Activism and Global Media: Perspective from the Past, Present and Future*, Cham: Springer Nature Switzerland, pp. 205–226.

Guerra, E. and Millard-Ball, A. (2017). Getting around a license-plate ban: Behavioral responses to Mexico City's driving restriction. *Transportation Research Part D: Transport and Environment, 55,* 113–126.

Kemasingki, P. (2020, January 30). Breathe... if you can. *Citylife Chiang Mai.* Retrieved on January 11, 2024 from https://www.chiangmaicitylife.com/clg/our-city/environment/breathe-if-you-can/.

KFEM. (n.d.). *Who we are.* Retrieved on January 11, 2024 from https://kfem.org/who-we-are/.

McNully, J. (2010). July 26, 1943: L.A. gets first big smog. *Wired.* Retrieved on January 11, 2024 from https://www.wired.com/2010/07/0726la-first-big-smog/.

Pan, W. (2017). Under the dome: Un-engineering digital capture in China's smog. *Asiascape: Digital Asia, 4*(1–2), 13–32.

Roberts, D. (2015, March 6). Opinion: How the US embassy tweeted to clear Beijing's air. *Wired.* Retrieved on January 11, 2024 from https://www.wired.com/2015/03/opinion-us-embassy-beijing-tweeted-clear-air/.

Turner, R. (2019, October 1). Smith Griswold sells the war against smog. *Science History Institute Museum & Library.* Retrieved on January 11, 2024 from https://www.sciencehistory.org/stories/magazine/smith-griswold-sells-the-war-against-smog/.

Chapter 13

Corporate–Civic Partnerships: Building Bridges for Azure Skies

Traditional corporate approaches to environmental management have often prioritized minimal compliance with regulations while maximizing shareholder value, reflecting our remarkable capacity for both rational calculation and moral compartmentalization. Corporate decision-makers, operating within systems that reward short-term financial metrics, often engage in the unconscious removal or dilution of ethical dimensions from decision-making processes. This narrow focus has historically led to the externalization of environmental costs onto communities, particularly in developing nations. The practice of "environmental dumping" exemplifies this, where corporations irresponsibly transfer hazardous waste or polluting operations to countries with weaker environmental regulations. For instance, the infamous case of British multinational Thor Chemicals during the 1980s and 1990s saw the company exporting toxic mercury waste to its Cato Ridge recycling plant in the KwaZulu-Natal province of South Africa. Apart from the occupational exposure of workers to the dangerous neurotoxin, investigations found that the company had discharged mercury waste into the local river systems.

The automotive industry has been particularly susceptible to this pattern of prioritizing profits over environmental responsibility, as stringent emissions regulations have often been viewed as obstacles to market competitiveness rather than necessary safeguards for public health and environmental protection. The case of Volkswagen's 2015 emissions scandal powerfully illustrates this dynamic. The company installed "defeat devices" in 11 million vehicles worldwide to circumvent emissions testing, prioritizing market share over environmental impact (Gates *et al.*, 2017). Oldenkamp *et al.* (2016) estimated that the fraudulent emissions were associated with "45 thousand disability-adjusted life years (DALYs) and a value of life lost of at least 39 billion US dollars" (p. 121). The company's initial cost–benefit analysis focused solely on financial metrics, exemplifying how traditional corporate thinking can systematically exclude ethical considerations. The subsequent billions of euros in fines and penalties demonstrated that even from a purely financial perspective, this approach was ultimately self-defeating. The scandal's aftermath sparked a broader industry-wide examination of emissions testing practices, leading to the discovery of similar devices in other automotive manufacturers' vehicles.

Yet some companies have successfully transformed their approach to environmental management. Consider Interface, the world's largest manufacturer of modular carpet tiles. Under the leadership of Ray Anderson, who experienced what he called a "spear in the chest" moment after reading Paul Hawken's *The Ecology of Commerce* in 1994, Interface embarked on "Mission Zero" — aiming to eliminate any negative environmental impact by 2020 (United Nations Climate Change, 2023). Anderson famously stated the following (Mason, 2011):

> I was dumbfounded by how much I did not know about the environment and about the impacts of the industrial system on the environment. A new definition of "success" began to creep into my consciousness, and the latent sense of legacy asserted itself. I was a plunderer of the Earth, and that is not the legacy one wants to leave behind.

Anderson's company achieved its goal through innovations like recycling fishing nets into carpet tiles via its Net-Works program (*PRNewswire*, 2015), significantly reducing water consumption, and operating with renewable electricity in manufacturing. Interface's market value grew from millions to several billion dollars during this transformation, demonstrating that environmental responsibility can align with financial success. The firm's success inspired other manufacturers to adopt similar practices. For example, Shaw Industries, another major flooring manufacturer, launched its own sustainability through an innovation initiative, guided by the cradle-to-cradle philosophy.

Even when specifically directed at resolving environmental problems, traditional corporate approaches embody linear, Newtonian thinking in addressing complex challenges leading to unintended consequences. Consider the electric vehicle revolution, heralded as a triumph of corporate innovation in combating urban air pollution. Tesla and other manufacturers operate from a mindset of technological solutionism, yet this approach overlooks the quantum nature of social impact. The celebrated growth of electric vehicle adoption created ripple effects that manifest as human rights violations in the cobalt mines of Congo, where communities face displacement, exploitation, and severe health consequences. This situation is further complicated by the geopolitical implications of resource control, with China securing significant influence over the global cobalt supply chain through strategic investments in Congolese mining operations. The environmental impact of lithium mining for EV batteries presents another challenge, particularly in the "lithium triangle" of South America (part of the Andes spanning the borders of Argentina, Bolivia, and Chile), where extraction processes threaten local water resources and Indigenous communities' livelihoods.

The complexity of these supply chain impacts becomes clear through specific metrics. The Democratic Republic of Congo supplies about 70% of the world's cobalt, a raw material needed for manufacturing batteries for electric vehicles (Beaule, 2023). Artisanal mining accounts for a significant part of this production, where thousands of miners — including children — work in hazardous conditions with no personal protective equipment. Beaule (2023) quoted researcher Siddharth Kara who described the situation in the mines as shocking and deplorable with "children caked in toxic grime and filth and scrounging in pits, trenches and tunnels to gather cobalt bearing ore and feed it up the supply chain." Kara (2023) further expounded on these human rights abuses related to the mining industry in his book *Cobalt Red*. The health impacts on these communities are severe and long-lasting, with studies showing elevated levels of cobalt and other heavy metals in blood samples from both miners and nearby residents, including children.

However, some companies are pioneering more responsible approaches to resource extraction. For instance, Fairphone, a Dutch social enterprise, has implemented a traceable supply chain for cobalt and other minerals. Their "Fair Cobalt Alliance" works directly with artisanal mining communities in the DRC to improve working conditions, eliminate child labor, and ensure fair compensation. While their market share remains small (selling approximately 90,000 phones annually), their approach demonstrates the feasibility of more ethical supply chain management. Relatedly, companies like BMW have committed to sourcing cobalt directly from mines in Morocco and Australia, where labor and environmental standards are relatively more strictly regulated.

Meanwhile, grassroots movements embody our deep-seated need for connection to place and community. These movements, powered by place attachment, often possess a rich understanding of local ecological relationships but lack the resources to implement large-scale solutions. Narratives like the Chinese film *Dying to Survive* based on real-life stories show such movements often emerge from profound moral courage and genuine community needs but frequently struggle with issues of scale, sustainability, and institutional legitimacy. The film powerfully illustrates struggle at the intersection of personal troubles and pressing public issues, demonstrating how individual acts of moral courage, while essential, need to be amplified through institutional channels to create lasting change.

The success of the Love Canal Homeowners Association (LCHA) in New York provides a compelling example of effective grassroots organizing. In 1978, Lois Gibbs, a local mother, discovered her children's school was built on top of a toxic waste dump. Despite initial dismissal from authorities, LCHA's persistent advocacy led to the evacuation of resident families and the passing of the Comprehensive Environmental Response, Compensation, and Liability Act, commonly known as the Superfund Act. The movement's success stemmed from combining compelling personal narratives with scientific evidence, as well as the courage to persist in picketing during the cold winter months.

The division between corporate and community approaches to tackling environmental challenges reflects a broader pattern in human society — our tendency to divide the world into "us" and "them," to create institutional structures that fragment what should be holistic approaches to complex problems. Yet the air we breathe acknowledges no such divisions. It flows freely between boardroom and backyard, factory and farmland, refusing to conform to our carefully constructed social categories.

The city of Pittsburgh's transformation illustrates both this division and potential paths to reconciliation. Described by James Parton in 1868 as "hell with the lid off" due to industrial pollution, Pittsburgh saw its air quality deteriorate to the point where streetlights needed to be lit at noon. The Donora Smog of 1948, which killed 20 people and sickened thousands, marked a turning point. This tragic event, occurring in the mill town of Donora, around 24 miles southeast of Pittsburgh, created a thick yellow smog that persisted for five days, causing respiratory distress in roughly half the town's population. The incident became one of the first documented air pollution disasters in American history and catalyzed the modern environmental movement. Today, Pittsburgh's air quality, while still concerning, has improved dramatically through multi-stakeholder collaboration. The *Breathe Project*, created by The Heinz Endowments in 2011, brings together corporate leaders, environmental scientists, and community advocates in a shared digital platform for air quality monitoring and improvement initiatives.

Recent years have witnessed the rise of more integrated approaches to air quality management. In Guangdong Province, China, several manufacturing facilities have implemented community-based air quality monitoring systems, giving local residents real-time access to emissions data and a voice in environmental decision-making. This transparency has not only improved trust but has also led to more effective pollution control measures. Similar initiatives have emerged in India, where textile manufacturers in Tamil Nadu have partnered with local environmental groups to develop green corridors around industrial zones. These projects combine modern industrial practices with traditional knowledge about local vegetation patterns, creating natural buffers that help filter air pollutants while preserving biodiversity. These partnerships have created boundary objects — concepts or things that have different meanings in different social worlds but enable coordination between them.

The emergence of effective corporate–civic partnerships requires what we might term "quantum social consciousness" — an awareness that corporate decisions create multiple sociological entanglements, that community impacts cannot be reduced to simple metrics, and that social systems exhibit nonlinear behavior where small interventions can have disproportionate effects. This quantum perspective reveals why traditional Corporate Social Responsibility initiatives often fail to have the expected impact — they attempt to artificially separate corporate activity from social impact when, in reality, they exist in a state of permanent entanglement. For example, the "one for one" model popularized by Toms' shoes, while seemingly beneficial, actually disrupted local shoe markets in some recipient communities, illustrating how linear thinking about social impact can produce unexpected negative consequences. We must recognize that every corporate decision creates multiple, simultaneous realities for different communities, that observation and participation are inseparable, and that social systems exhibit properties of non-local causality that demand a more sophisticated approach to impact assessment and intervention.

Analysis of effective corporate–civic partnerships reveals several crucial elements that contribute to their success. Transparent accountability mechanisms, established through clear metrics for air quality improvement and regular public reporting of progress, help maintain trust and ensure all parties remain committed to environmental goals. Rather than treating community engagement as a box-checking exercise, effective partnerships integrate local voices into decision-making processes from the outset, including representation on environmental management committees and regular community consultations. The most successful partnerships recognize the value of both technical expertise and local ecological knowledge. For instance, several industrial facilities have improved their environmental management systems by incorporating the traditional understanding of wind patterns and natural ventilation principles. Sustainable partnerships extend beyond short-term public relations gains to establish enduring mechanisms for collaboration, including dedicated funding streams for environmental monitoring and community capacity building.

Despite their potential, corporate–civic partnerships face several challenges. Power imbalances between large corporations and local communities can skew decision-making. Limited technical capacity within community groups can hinder meaningful participation. Many grassroots organizations lack access to environmental monitoring equipment, data analysis capabilities, or legal expertise necessary for effective advocacy. Additionally, maintaining long-term commitment from both parties requires careful attention to incentive structures. The history of environmental agreements is littered with examples of partnerships that dissolved once public attention waned or economic conditions changed. Successful partnerships have addressed these challenges through independent facilitation by neutral third parties, technical capacity-building programs for community members, formal agreements that outline rights and responsibilities, and regular review and adjustment of partnership frameworks.

While some challenges persist, recent developments have introduced new tools and frameworks for addressing them. The concept of environmental personhood, which grants legal rights to natural entities, has gained traction globally. In March 2017, after a long legal battle, the Whanganui River in New Zealand became the Earth's first river to be given the same legal rights as an individual (Hollingsworth, 2020). This implies that any development projects concerning the Whanganui, whether government or corporate-led, would need to seriously consider the river's health and well-being. A few days later, an Indian court in Uttarakhand declared the Yamuna and Ganges rivers to be "legal and living entities," justifying that the dire pollution situation in India required "extraordinary measures to be taken to preserve and conserve" these sacred rivers (Suri, 2017). However, the Supreme Court of India later overturned this ruling by stating that the declaration was legally unsustainable (*BBC*, 2017). This legal framework has important implications for addressing air pollution, as it opens the possibility of granting legal rights to the atmosphere itself. Recognizing air as a legal entity could strengthen enforcement of air quality standards and create new mechanisms for holding polluters accountable. While such proposals face significant political and legal hurdles, they represent an innovative approach to reinforcing existing environmental regulations and creating new pathways for protecting air quality.

The evolution of legal frameworks like environmental personhood demonstrates how fundamental changes in our conceptual approach can create new possibilities for environmental protection. Just as granting legal rights to natural entities transforms our relationship with the environment, we must similarly reimagine the dynamics between corporations and communities. Successful corporate–civic partnerships require a fundamental shift in how we conceptualize the relationship between business and community. Rather than viewing community engagement as a constraint on corporate activity, companies need to recognize it as a source of innovation and competitive advantage. This would reflect a growing understanding that environmental challenges cannot be effectively addressed through traditional corporate or community approaches alone, but require the integration of diverse perspectives, knowledge systems, and capabilities.

Corporate–civic partnerships suggest a path forward that neither demonizes industry nor romanticizes traditional approaches, but instead seeks to combine the strengths of both corporate and community stakeholders. As we face increasingly complex environmental challenges, such collaborative approaches will become ever more crucial for achieving and maintaining clearer skies for all. Looking ahead, several emerging trends promise to reshape corporate–civic environmental partnerships. The concept of "regenerative business" that focuses on actively improving the health of the entire ecosystem is gaining traction, moving beyond sustainability to restore environmental and social systems (Hahn and Tampe, 2021). Similarly, the B Corp movement is creating new legal and social frameworks for businesses that explicitly recognize community and environmental stakeholders. These developments suggest that the future of corporate–civic partnerships will require even greater attention to social-ecological resilience — the capacity of linked social and environmental systems to adapt and thrive in the face of change.

References

BBC. (2017, July 7). India's Ganges and Yamuna rivers are 'not living entities'. *BBC News*. Retrieved on April 15, 2024 from https://www.bbc.com/news/world-asia-india-40537701.

Beaule, V. (2023, February 8). Artisanal cobalt mining swallowing city in Democratic Republic of the Congo, satellite imagery shows. *ABC News*. Retrieved on April 15, 2024 from https://abcnews.go.com/International/cobalt-mining-transforms-city-democratic-republic-congo-satellite/story?id=96795773.

Gates, G., Ewing, J., Russel, K., and Watkins, D. (2017, March 16). How Volkswagen's 'Defeat Devices' worked. *The New York Times*. Retrieved on February 2, 2024 from https://www.nytimes.com/interactive/2015/business/international/vw-diesel-emissions-scandal-explained.html.

Hahn, T. and Tampe, M. (2021). Strategies for regenerative business. *Strategic Organization*, *19*(3), 456–477.

Hollingsworth, J. (2020, December 11). This river in New Zealand is legally a person. Here's how it happened. *CNN*. Retrieved on February 2, 2024 from https://www.cnn.com/2020/12/11/asia/whanganui-river-new-zealand-intl-hnk-dst/index.html.

Kara, S. (2023). *Cobalt Red: How the Blood of the Congo Powers Our Lives*. New York: St. Martin's Press.

Mason, P. (2011, August 26). Ray Anderson obituary. *The Guardian*. Retrieved on February 2, 2024 from https://www.theguardian.com/environment/2011/aug/26/ray-anderson-obituary.

Oldenkamp, R., van Zelm, R., and Huijbregts, M. A. (2016). Valuing the human health damage caused by the fraud of Volkswagen. *Environmental Pollution*, *212*, 121–127.

PRNewswire. (2015, June 10). Net-Works: The world's first inclusive business model to recycle discarded fishing nets Made in the Philippines and now primed to go global. *Interface*. Retrieved on February 2, 2024 from https://

investors.interface.com/news/press-release-details/2015/Net-Works-The-worlds-first-inclusive-business-model-to-recycle-discarded-fishing-nets-Made-in-the-Philippines-and-now-primed-to-go-global/default.aspx.

Suri, M. (2017, March 22). India becomes second country to give rivers human status. *CNN*. Retrieved on February 2, 2024 from https://edition.cnn.com/2017/03/22/asia/india-river-human/index.html.

United Nations Climate Change. (2023). From Mission Zero to climate take back: How Interface is transforming its business to have zero negative impact. Retrieved on February 2, 2024 from https://unfccc.int/climate-action/momentum-for-change/climate-neutral-now/interface.

Chapter 14

Flight to B Economy: Redefining Corporate Success Through ESG Metrics

In September 2022, Yvon Chouinard made a decision that sent tremors through corporate boardrooms worldwide. The founder of Patagonia, a company valued at $3 billion, walked into his office and signed away his family's ownership of the business. Not to a private equity firm. Not to a competitor. Not even to his children. Instead, he transferred the company's voting stock to a purpose-built trust and its remaining shares to Patagonia Purpose Trust and the Holdfast Collective. His mandate? Use Patagonia's profits — the part that's not reinvested back into the business — to combat climate change and protect undeveloped land around the globe.

"Earth is now our only shareholder," Chouinard announced in an open letter that for some, seemed like corporate suicide, and for others, represented the dawn of a new economic paradigm. Perhaps for a billionaire to give away his company seemed unprecedented, but even more remarkable was Chouinard's reasoning: "Instead of extracting value from nature and transforming it into wealth, we are using the wealth Patagonia creates to protect the source" (Patagonia, 2022). The company would continue operating as a for-profit business, but with a radically different purpose. This transformation was made possible by Patagonia's status as a B Corporation — a new kind of company that legally commits to considering environmental and social impacts alongside profits.

While the previous chapter illuminated how corporate–civic partnerships can bridge the divide between business interests and community needs, this chapter sheds light on a more fundamental transformation — the emergence of a new economic paradigm that redefines corporate success itself. The transition from traditional profit-centric metrics to comprehensive Environmental, Social, and Governance (ESG) frameworks represents more than just an evolution in measurement — it embodies what the previous chapter termed "quantum social consciousness" in corporate thinking. This shift has given rise to a momentum to build the B Economy that "supports businesses striving to create a shared and durable prosperity for all" (B the Change, 2018). The move toward B Economy is driven by mission-driven businesses like certified benefit corporations. Unlike traditional for-profit firms, these legally incorporated benefit corporations are required to create public benefit alongside financial returns.

The transformation toward a B Economy marks a decisive break from conventional business models that have dominated the past century. These traditional approaches followed a predominantly linear trajectory of taking resources, making products, and ultimately disposing of waste — typically through landfills, incineration, or ocean dumping. While this take-make-waste pathway has undeniably fueled immense prosperity and lifted millions out of poverty through mass production and consumption, it has also externalized monumental environmental costs in its pursuit of maximizing financial gains. The mounting challenges of runaway greenhouse gas emissions, biodiversity destruction, and pollution now pose existential threats to both ecological stability and human well-being.

The transition to a B Economy necessitates moving beyond conventional corporate accounting that has historically revolved around financial statements like income statements, cash flow statements, and balance sheets. This traditional focus on profitability and asset valuation captures only a fragment of company performance, disproportionately emphasizing economic metrics over sustainability impacts. As a result, major costs passed to society and the environment get excluded from decision-making calculus. To illustrate, despite 83% of leading CEOs voicing support for urgent climate action, 50% admit financial targets currently override environmental ones at their firms (UN Global Compact and Accenture, 2021). This ethos sees upstream supply chain impacts ignored while waste and pollution remain unchecked so long as growth continues.

The transformation toward a B Economy is perhaps best understood through the concrete examples of corporate failure and success in addressing societal challenges. The stark contrast between Volkswagen's emissions scandal in 2015 and today's emerging purpose-driven enterprises highlights how profoundly business values have evolved over the recent decade. Where Volkswagen exemplified the old paradigm's narrow focus on financial metrics and shareholder value, companies presently pursuing B Corp certification represent a new understanding that corporate decisions create multiple, simultaneous realities for different communities. For example, consider Interface, a company that dared to imagine a different path. Their "Mission Zero" wasn't just about environmental improvements — it pioneered a new way of measuring and defining corporate success. Its development of modular carpet tiles that could be individually replaced reduced both waste and installation costs. Its biomimicry-inspired design processes led to breakthrough products that consumed less energy and materials. These innovations didn't just benefit the environment — they created competitive advantages that transformed Interface into an industry leader. The company's market value growth from millions to billions while pursuing aggressive environmental goals demonstrated how ESG metrics could align financial success with environmental stewardship. This shift mirrors the previous chapter's observation that social systems exhibit nonlinear behavior where small interventions can have disproportionate effects.

The B Economy movement's potential impact on air quality illustrates this nonlinear relationship between business transformation and environmental outcomes. Air pollution, a complex challenge that transcends national boundaries and economic sectors, requires precisely the kind of systemic rethinking that B Corps champion. Traditional approaches to air quality management have often focused on end-of-pipe solutions and regulatory compliance, treating emissions as an inevitable byproduct of economic activity. However, B Corps are demonstrating that reimagining business models can address air pollution at its source while creating new value opportunities. For instance, when a certified B Corp commits to powering its operations with 100% renewable energy, it not only reduces its direct emissions but also catalyzes change throughout its supply chain. Its suppliers, facing pressure to align with these standards, will likely begin their own transitions to cleaner energy sources, creating a multiplier effect that extends far beyond the company's immediate operations.

The emergence of legalized B Corps represents the formalization of a new approach to business. Unlike traditional corporate social responsibility initiatives that often attempt to artificially separate corporate activity aimed at shareholder primacy from social impact, now countries and states can pass the necessary legislation to allow new companies to incorporate as "benefit corporations" and existing companies to amend their governing documents to become benefit corporations. In addition, governments could offer tax benefits for B Corps to potentially offset the costs of certification and ongoing compliance, while procurement preferences in government contracts could create market incentives for certification. The development of specific legal structures might help protect companies pursuing multiple bottom lines. This legal evolution also reflects a growing recognition that taken-for-granted accounting practices disproportionately stressing financial performance have inhibited sustainable business transformation. Conventional reporting fails to capture major externalities that accumulate across product lifecycles from raw material extraction through disposal. Quantifying and embedding such environmental, health, and social costs are essential for systemic change that aligns commercial success with ecological viability.

Several states in the US have already passed related legislation, and there are significant differences in these statutes, which help to "create a laboratory within which different approaches to transparency, reporting standards, accountability, and director/officer fiduciary duties to stakeholders can be studied and compared" (Hiller, 2013, p. 299). In 2015, Italy became among the first countries outside the US to pass the relevant law known as *Società Benefit*. Del Baldo (2019) notes that B Corps not only legally commit to creating "a material positive impact on society and the environment" (p. 2) but also need to expand the duties of the Board of Directors to consider the interests of non-financial stakeholders. In addition, B Corps have an obligation to report on their overall ESG performance using "a comprehensive, credible, independent and transparent third-party standard" (Del Baldo, 2019, p. 2).

Different types of ESG metrics provide the measurement framework for B Corps. B Lab, a non-profit founded in 2006 spearheading the B Corp movement, created the *B Impact Assessment* as a freely available tool to measure corporate performance along different ESG dimensions (B Lab, n.d.). These metrics measure not just direct environmental impacts but also assess companies' relationships with communities, their governance structures, and their long-term sustainability strategies. For instance, environmental metrics now go beyond simple emission measurements to evaluate companies' entire ecological footprint, including their impact on biodiversity, water resources, and air quality. Social metrics examine workforce practices, community relationships, and supply chain responsibility. Governance metrics assess companies' decision-making structures, transparency, and stakeholder engagement practices.

That said, the complexity of global supply chains, the difficulty of quantifying social impact, and the need for standardized measurement frameworks remain significant challenges. However, innovative solutions are emerging. The development of blockchain-based supply chain tracking systems allows companies to verify their environmental and social impact claims. Artificial intelligence and big data analytics also help companies process complex ESG data and identify areas for improvement. These technological tools enable more accurate and comprehensive ESG reporting. Additionally, the COVID-19 pandemic accelerated the shift toward a B Economy by highlighting the interconnectedness of business success and societal well-being. Companies with strong ESG practices generally showed greater resilience during the crisis, demonstrating the practical value of stakeholder-focused business models. The pandemic also exposed the vulnerabilities of global supply chains, leading many companies to reconsider their relationship with local communities and environments.

The transition to B Corp status offers compelling advantages. Modern consumers increasingly favor companies that demonstrate an authentic commitment to social and environmental causes, making B Corp certification a valuable third-party validation of such commitments (Bianchi *et al.*, 2020). Furthermore, with Millennial and Gen Z workers strongly preferring employers whose values align with their own, B Corp status serves as a powerful talent attraction and retention tool. In addition, the comprehensive *B Impact Assessment* used to measure progress toward B Corp certification can help reveal operational inefficiencies and risks that might otherwise go unnoticed, leading to more robust business models. However, a more fundamental change in dominant institutional logics is needed for widespread B Economy adoption. Overcoming the deeply entrenched taken-for-granted shareholder primacy mindset requires not just new legal structures and measurement frameworks, but a fundamental shift in how we conceive business success. Building trust in new business models and educating consumers about B Corp's significance demand sustained effort from multiple stakeholders. For example, supporting business education programs that emphasize stakeholder capitalism and sustainable business practices could also help create a pipeline of leaders equipped to navigate this new economic paradigm. In this regard, the Association to Advance Collegiate Schools of Business has taken a strong step by making "societal impact" a part of its 2020 accreditation standards (AACSB, n.d.).

The B Economy's potential becomes particularly evident when considering air pollution challenges in developing nations, where rapid industrialization often creates a perceived tension between economic growth and environmental protection. Traditional economic models have perpetuated the false choice between prosperity and clean air. However, B Corps are demonstrating how businesses can help break this paradigm. When B Corps implement cleaner production processes and share their technological innovations with smaller suppliers throughout their networks in various geographical locations, such knowledge transfer helps democratize access to cleaner technologies in developing economies.

However, emerging research suggests that B Corp certification alone may not guarantee meaningful organizational transformation. A study of B Corps in Brazil revealed that many companies maintained their existing governance structures post-certification, with limited expansion of stakeholder engagement beyond shareholders (Villela *et al.*, 2021). According to the authors' analysis, despite achieving high initial certification scores, several companies failed to set improvement goals between certifications or make substantial progress in their impact scores over time. This suggests that some B Corps may fall into a "certification trap" — using the designation primarily for reputational benefits while avoiding deeper structural changes. As Winkler *et al.* (2018) argued, more distributed governance models, such as employee ownership schemes, could help address this limitation by embedding stakeholder engagement more deeply into organizational structure. This challenge echoes a familiar pattern in corporate sustainability initiatives, where surface-level compliance can sometimes substitute for substantive change. The transition to a true B Economy may therefore depend not just on certification frameworks but on the emergence of leadership willing to fundamentally reimagine corporate governance and stakeholder relationships.

Perhaps this tension reflects a deeper truth about human nature — our capacity to simultaneously recognize profound problems while continuing to engage with them imperfectly. Like viewing a mountain from different angles, the path to corporate responsibility appears different depending on one's perspective and position. Some may see B Corp certification as a summit to be reached, while others understand it as merely one ridge in a longer journey. The challenge lies not in achieving perfection, but in maintaining forward momentum despite our limitations, knowing that each life affected by corporate actions — whether through environmental impact or social change — holds both infinite value and everyday fragility.

Looking ahead, the B Economy represents more than just a new way of measuring corporate success — it offers a framework for addressing the challenges identified throughout this book. By redefining success to include environmental and social metrics alongside financial ones, it creates institutional structures that support rather than undermine environmental stewardship. This transformation suggests that the path to clearer skies lies not just in better technology or stricter regulations, but in fundamentally reimagining how we define and measure corporate success. As we face mounting environmental and social challenges, the B Economy offers a promising path forward, one that recognizes the inherent interconnectedness of business success and societal well-being and provides practical frameworks for implementing the theoretical insights gained from studying corporate–civic partnerships and stakeholder engagement. It is certainly not a panacea, but one of many necessary steps in humanity's ongoing struggle to align commercial activity with deeper values. The story of Yvon Chouinard and Patagonia with which we started this chapter represents not an end point but a beginning. As more companies follow this path, we may find that the B Economy isn't just a better way of doing business — it might be one of the key steps to ensuring our collective survival and prosperity on a finite planet.

References

AACSB. (n.d.). Societal impact at AACSB. *Association to Advance Collegiate Schools of Business*. Retrieved on September 29, 2024 from https://www.aacsb.edu/about-us/advocacy/societal-impact.

Bianchi, C., Reyes, V., and Devenin, V. (2020). Consumer motivations to purchase from benefit corporations (B Corps). *Corporate Social Responsibility and Environmental Management, 27*(3), 1445–1453.

B Lab. (n.d.). B Impact Assessment. Retrieved on September 20, 2024 from https://www.bcorporation.net/en-us/programs-and-tools/b-impact-assessment/.

B the Change. (2018, September 30). 5 Ways we are building the B Economy. *Medium.* Retrieved on February 12, 2024 from https://bthechange.com/4-ways-we-are-building-the-b-economy-d0e2958c15fb.

Del Baldo, M. (2019). Acting as a benefit corporation and a B Corp to responsibly pursue private and public benefits. The case of Paradisi Srl (Italy). *International Journal of Corporate Social Responsibility, 4*, 1–18.

Hiller, J. S. (2013). The benefit corporation and corporate social responsibility. *Journal of Business Ethics, 118,* 287–301.

Patagonia. (2022, September 14). Patagonia's next chapter: Earth is now our only shareholder. Retrieved on February 12, 2024 from https://www.patagonia-works.com.

UN Global Compact and Accenture. (2021, November 1). The 2021 United Nations Global Compact — Accenture CEO Sustainability Study. Retrieved on February 12, 2024 from https://unglobalcompact.org/library/5976.

Villela, M., Bulgacov, S., and Morgan, G. (2021). B Corp certification and its impact on organizations over time. *Journal of Business Ethics, 170,* 343–357.

Winkler, A. L. P., Brown, J. A., and Finegold, D. L. (2019). Employees as conduits for effective stakeholder engagement: An example from B corporations. *Journal of Business Ethics, 160,* 913–936.

Chapter 15

The Right to Breathe: Air Quality as the Next Human Rights Frontier

In December 2020, Ella Adoo-Kissi-Debrah made legal history. A London coroner ruled that air pollution had made a "material contribution" to her death, marking the first time in the world that air pollution was officially listed as a cause of death on a death certificate (London Inner South Coroner's Court, 2020). Ella had lived near London's South Circular Road, a pollution hotspot where nitrogen dioxide levels from traffic emissions consistently exceeded World Health Organization (WHO) guidelines. After she died at nine years of age from acute respiratory failure in February 2013, the Court's verdict originally did not investigate the issue of air pollution. Her mother's subsequent tireless campaign not only highlighted the deadly impact of toxic air on children but also fundamentally reframed the issue — breathing clean air wasn't just an environmental concern, but a basic human right. In 2024, the Mayor of London, Sadiq Khan, delivered an apology to Ella's family on behalf of the Greater London Authority (Mayor of London, 2024):

> Sorry. You deserved so much better. But — like you — I'm determined to keep Ella's memory and legacy alive by ensuring that we do not repeat the mistakes of the past by failing today's generation of young Londoners (...) I'll also continue putting pressure on the Government to do more across the country to prevent further needless suffering and death, including by adopting 'Ella's law'. This would make breathing clean air what it should be — a right that must be upheld.

Here, "Ella's law" refers to the Clean Air (Human Rights) Bill, introduced to the United Kingdom's Parliament by Baroness Jenny Jones in May 2022. The bill would have forced "the government to act to bring air quality in every community up to minimum WHO standards" within the next five years (The Ella Roberta Foundation, n.d.). Although the bill was later rejected by the government in 2023, it embodies an important attempt to ensure that every child in the UK could breathe clean air, regardless of their ethnic background or their economic status.

The recognition of clean air as a human right has profound implications for business operations worldwide. Companies are increasingly finding themselves at the intersection of human rights law and environmental regulations, facing new forms of liability and accountability. The story of the Swiss conglomerate Holcim (which previously merged with the French company Lafarge) serves as a harbinger of this shift. In 2022, the cement giant, which was already embattled by allegations of aiding and abetting crimes against humanity due to Lafarge's operations in Syria (Alderman, 2018), faced a landmark legal challenge when four inhabitants of the Indonesian island of Pari sued the company for climate change-related damages in their community. *Asmania et al. vs Holcim* represents "the first time a Swiss court addresses the question of whether a corporation can be held liable under civil law for its contribution to climate change" (European Center for Constitutional and Human Rights, 2023). The plaintiffs are demanding that Holcim "immediately and significantly reduce its carbon dioxide emissions, pay compensation for damages already incurred, and co-finance urgently needed flood protection measures" in Pari (HEKS/EPER, 2022).

The business rationale for supporting the right to clean air extends far beyond mere regulatory compliance or corporate social responsibility. Companies are discovering that their long-term viability increasingly depends on how they address air quality issues. When examining potential risks, corporations face a complex web of interconnected challenges. Legal exposure has grown dramatically as more jurisdictions recognize clean air rights — companies now confront not just regulatory fines but potentially massive litigation costs. The reputational stakes have also escalated; in an era where consumers can instantly access air quality data and track corporate environmental impact, companies found to be significant polluters face substantial public backlash.

The repercussions of poor corporate environmental stewardship can be swift and severe, as demonstrated by a dramatic incident in December

2022. When approximately 200 environmental activists stormed a Lafarge (part of the Holcim conglomerate group) cement factory in Southern France, their actions — from damaging construction equipment and sabotaging incinerators to opening bags of cement and spraying graffiti — highlighted the growing public intolerance for corporate environmental negligence. The responsible group named "Action Lafarge" explained their motivation (Colwell, 2022):

> The Lafarge Holcim group, with its billions of turnover, does not shrink from anything to continue its insane race for profit, and this in disregard of all the ecological and social consequences generated (...) Here in Bouc-Bel-Air the company has never hesitated to lobby to exceed the environmental standards for dust and sulfur oxides set by the European Union. Of the 50 most polluting sites in France, 20 are cement plants, including this one, which produces more than 444,464 tons of CO_2 per year and feeds its kilns with thousands of old tires and all kinds of toxic waste.

The incident served as a stark reminder that companies can no longer simply manage environmental issues through public relations or minimal compliance efforts.

The transformation of corporate attitudes toward air quality rights is perhaps best illustrated through the evolution of industry pioneers. Interface's journey from traditional carpet manufacturer to environmental leader demonstrates how companies can fundamentally reimagine their relationship with air quality. Faced with the reality that traditional carpet manufacturing released significant volatile organic compounds (VOCs), Interface didn't simply aim to meet existing standards. Instead, it invested heavily in research and development, eventually creating new manufacturing processes that virtually eliminated VOC emissions (Interface, n.d.). What makes Interface's story particularly noteworthy is its decision to share these innovations with competitors. This choice reflected a profound understanding that the right to breathe clean air transcends traditional competitive boundaries. By openly sharing its technological breakthroughs, Interface helped elevate entire industry standards while demonstrating that environmental leadership and business success can go hand in hand.

The legal landscape surrounding air quality rights continues to evolve rapidly, creating new challenges and opportunities for corporate entities.

The Urgenda Foundation case in the Netherlands served as a watershed moment, establishing government obligations to protect citizens from climate change and air pollution. In *Urgenda Foundation v. State of the Netherlands*, originally filed in 2015, the Urgenda Foundation, along with 900 Dutch citizens, sued the Dutch government to require it to do more to prevent global climate change. After multiple appeals, in December 2019, the Supreme Court of the Netherlands upheld the lower court's decision requiring the Dutch state to limit greenhouse gas emissions to 25% below 1990 levels by 2020 (de Wit, 2019). This decision has reverberated through corporate boardrooms worldwide, as companies realize they could increasingly face similar litigation alongside government entities.

South Korea's *Act on Liability for Environmental Damage and Relief Thereof* enacted in 2014 offers another compelling example of how legal frameworks are evolving. The Act aimed to provide the victims of environmental damage with prompt relief instead of having to engage in long expensive litigations by reducing the victim's burden of proof. This made it an obligation for business owners "to have relssevant environmental liability insurance, presumption of causal relationship, the right to request information, and payment of relief money to the victims of environmental damage" (Lee and Ko, 2019). The implications are profound — companies must now consider potential health impacts more proactively or face significant legal consequences.

Moreover, in August 2024, South Korea's Constitutional Court issued a first-of-its-kind decision in Asia by ruling unanimously that the provision on quantitative greenhouse gas emission reduction targets specified in the country's 2021 *Framework Act on Carbon Neutrality and Green Growth for Coping with Climate Crisis* was unconstitutional because it failed "to protect the rights of future generations and must be amended by February 2026" (Davies *et al.*, 2024). The Court found that the reduction targets shifted an excessive burden to the future, and thus lacked the features of a minimum protective measure against the risks of climate change. It prescribed that the government create year-by-year carbon reduction targets for 2031 to 2049.

Apart from the legal landscape, the financial sector has also emerged as a powerful force in advancing the right to clean air, creating economic incentives for corporate environmental responsibility. The World Bank's new Environmental and Social Framework explicitly recognizes air quality impacts as a human rights consideration in project funding. Similarly, major investment firms like BlackRock now include air quality metrics in their ESG assessments, creating financial incentives for corporations to protect air quality. This shift in financial priorities has been accompanied by the emergence of new corporate structures, such as Benefit Corporations. Unlike traditional corporate structures that primarily prioritize shareholder value, B Corps legally commit to considering their impact on all stakeholders, including the environment and affected communities. Companies like Patagonia, which became a B Corp in 2012, demonstrate how this new corporate structure enables a more comprehensive approach to environmental stewardship while maintaining business success. Similarly, firms like Danone North America, the largest B Corp in the world, demonstrate how businesses can evolve their strategies and reward systems to protect environmental rights while maintaining business viability. Their approach to air quality includes comprehensive monitoring systems, transparent reporting mechanisms, and integration of air quality metrics into executive compensation structures.

This structural evolution in corporate governance has inspired even traditional companies to reimagine their approach to air quality. Microsoft's ambitious commitment to becoming carbon negative by 2030 exemplifies how corporations can leverage their market power to drive systemic change (Smith, 2020). Rather than limiting its focus to direct emissions, Microsoft has extended its influence throughout its supply chain, requiring suppliers to report and reduce their air quality impacts. This approach recognizes that meaningful protection of air quality rights requires addressing the entire ecosystem of corporate relationships.

Corporate innovation in addressing air quality challenges continues to evolve through strategic partnerships and technological advancement. L'Oréal's multi-year partnership with climate tech company BreezoMeter demonstrates how companies can integrate the values of science and innovation into their environmental strategies. By partnering with the world air quality leader BreezoMeter, L'Oréal intends to unpack the relationships "between skin aging and environmental exposures, such as allergens, UV, and pollution" (Rydzek, 2022). Their approach will enable L'Oréal to develop more precise and personalized skincare products suitable for customers facing different types of environmental hazards. Relatedly, Tesla's development of HVAC systems shows how companies can leverage their core technological expertise to address broader air quality challenges. Tesla also opened its electric vehicle patents to competitors, demonstrating how companies can prioritize environmental progress over short-term competitive advantage. These initiatives demonstrate that companies can find new business opportunities in environmental protection. Such partnerships, innovations, and sharing practices indicate a growing recognition that environmental protection and business success are increasingly intertwined.

The acknowledgment of access to clean air as a basic human right demands a fundamental transformation in corporate behavior and strategy. Companies need to begin by embedding this right into their governance structures, moving beyond traditional environmental policies to explicitly acknowledge clean air as a fundamental human right. This requires more than just updating corporate documents — it means creating accountability mechanisms and integrating air quality considerations into all major business decisions. It also implies moving beyond minimal compliance reporting to embrace radical transparency about their air quality impacts. This could include real-time monitoring, publicly accessible data platforms, and detailed reporting on both direct and indirect emissions. Such transparency not only builds trust with stakeholders but also creates internal pressure for continuous improvement.

In addition, investment in clean technologies and processes has to be approached as a strategic imperative rather than a regulatory burden. Companies should view such investments through the lens of innovation opportunity, recognizing that solutions to air quality challenges often drive broader technological advancement and competitive advantage. This mindset shift would help align environmental protection with business growth. Companies that embrace this shift — moving beyond compliance to become active protectors of air quality — will likely find themselves better positioned for long-term success. They'll face fewer legal and reputational risks, stronger community relationships, and better access to capital markets that increasingly value environmental performance.

Companies also need to engage more meaningfully with affected communities, viewing them not as potential adversaries but as essential partners in protecting air quality rights. This involves creating meaningful dialogue mechanisms, sharing decision-making power, and ensuring that communities benefit from corporate air quality initiatives. It could also include advocacy work for clear regulatory frameworks that protect relevant communities' air quality rights.

As Ella Adoo-Kissi-Debrah's case showed, the human cost of air pollution is too high to ignore, and businesses have a crucial role to play in protecting the right to breathe clean air. Her tragic death illuminated the stark contradictions that plague our modern society — the tension between corporate prosperity and human welfare, between technological progress and environmental preservation, and between immediate profit and long-term survival. Through her sacrifice, Ella became a beacon of change, her story a testament to how a life lost can illuminate the path forward for millions. The path toward clean air is not singular but manifold, with countless approaches waiting to be discovered and implemented — from technological innovation and policy reform to community engagement and corporate restructuring. Each path offers its own promise, its own challenges, and its own potential for meaningful change. The question today is no longer whether clean air is a human right, but how quickly and effectively the corporate world will adapt to help protect it.

References

Alderman, L. (2018, June 28). French cement giant Lafarge indicted on terror financing charge in Syria. *The New York Times.* Retrieved on October 7, 2024 from https://www.nytimes.com/2018/06/28/business/lafarge-holcim-syria-terrorist-financing.html.

Colwell, P. (2022, December 12). 200 French environmental activists Sabotage Lafarge-Holcim Marseille cement plant. *Atlas News.* Retrieved on October 7, 2024 from https://theatlasnews.co/business/2022/12/12/200-french-environmental-activists-sabotage-lafarge-holcim-marseille-cement-plant.

Davies, P., Kang, W., Green, M., Bee, J., Forrest, S., and Park, J. (2024, September). Ruling requires South Korean government to review climate targets. *Latham & Watkins LLP.* Retrieved on October 7, 2024 from https://www.globalelr.com.

de Wit, E. (2019). Urgenda Foundation v Netherlands: Historic climate change decision upheld. *Norton Rose Fulbright LLP.* Retrieved on October 7, 2024 from https://www.nortonrosefulbright.com/en-ca/knowledge/publications/45dc4f83/urgenda-foundation-v-netherlands-historic-climate-change-decision-upheld.

European Center for Constitutional and Human Rights. (2023). Climate litigation against Holcim: The four plaintiffs are granted legal aid. *ECCHR Press Releases.* Retrieved on November 14, 2024 from https://www.ecchr.eu/.

HEKS/EPER. (2022, July 12). *An island demands justice.* Dossier for the press conference on "Europe's economic responsibility for the climate crisis: The case of Pari Island, Indonesia" in July 2022. Retrieved on October 7, 2024 from https://callforclimatejustice.org/en/webreport/.

Interface. (n.d.). The proof behind our products. Retrieved on October 7, 2024 from https://www.interface.com/US/en-US/sustainability/certifications-documentation.

Lee & Ko. (2019). Environmental law in South Korea. *Lexology*. Retrieved on October 7, 2024 from https://www.lexology.com/.

London Inner South Coroner's Court. (2020). Inquest touching the death of Ella Roberta Adoo Kissi-Debrah. Record of Inquest. Retrieved on October 7, 2024 from https://www.innersouthlondoncoroner.org.uk/news/2020/nov/ inquest-touching-the-death-of-ella-roberta-adoo-kissi-debrah.

Mayor of London. (2024, February 2). Mayor delivers apology on behalf of GLA for death of Ella Adoo-Kissi-Debrah. Retrieved on October 7, 2024 from https://www.london.gov.uk/.

Rydzek, C. (2022, January 7). L'Oréal announced a multi-year partnership with climate tech company Breezometer. *TheIndustry.beauty*. Retrieved on October 7, 2024 from https://theindustry.beauty/loreal-announced-a-multi-year-partnership-with-climate-tech-company-breezometer/.

Smith, B. (2020, January 16). Microsoft will be carbon negative by 2030. *Official Microsoft Blog*. Retrieved on October 7, 2024 from https://blogs.microsoft.com/blog/2020/01/16/microsoft-will-be-carbon-negative-by-2030/.

The Ella Roberta Foundation. (n.d.). *Ella's Law*. Retrieved on October 7, 2024 from https://www.ellaroberta.org/campaigns/ellas-law.

Epilogue: Shaping Blue Skies for All

The world is a dangerous place, not because of those who do evil, but because of those who look on and do nothing — Einstein

As we grapple with the global challenge of air pollution, particularly acute in developing nations like India, we are confronted not just with technological and policy hurdles, but with the very essence of human nature. The concept of the "privilege of blue skies" is a stark reminder of global inequalities in environmental quality. In cities like Copenhagen, Oslo, or Wellington, residents wake up to crystalline skies and take for granted the simple act of breathing clean air. Meanwhile, in Delhi, Lahore, or Kinshasa, millions start their day checking air quality apps as routinely as weather forecasts, their children often unable to play outdoors for weeks at a time during pollution peaks. At the heart of our struggle lies a fundamental aspect that motivates human beings: self-interest. This trait, which has driven much of human progress and innovation, also serves as a significant obstacle in our collective efforts to address global environmental challenges. In developed countries, individuals have grown accustomed to the quality of life afforded by mass consumption, a lifestyle that often comes at the cost of environmental degradation elsewhere. The ability to enjoy clean air while the pollution-intensive industries that support this lifestyle are offshored to developing nations creates a psychological distance from the problem, allowing for a form of environmental cognitive dissonance. This phenomenon is particularly evident in the textile industry, where major fashion brands have relocated manufacturing to countries like Bangladesh and Vietnam, leading to severe air and water pollution in these regions while maintaining pristine retail environments in Western nations.

Conversely, in middle- and low-income countries, the promise of economic growth and poverty alleviation through industrialization presents an alluring path that seems to justify the temporary sacrifice of environmental quality. The immediate tangible benefits of job creation and increased standard of living often outweigh the less visible, long-term costs of air pollution. This prioritization is not merely a policy choice but a reflection of the human tendency to value immediate gains over future benefits, a phenomenon well documented in behavioral economics and psychology. The philosopher Peter Singer (1981), in his work on expanding the circle of ethical consideration, argues that human moral progress has been characterized by an ever-widening sphere of concern. From caring only for ourselves and our immediate family, we have gradually extended our moral consideration to our tribe, our nation, and in some cases to all of humanity. The challenge now is to extend this circle further, to include not just all currently living humans but future generations and the environment itself. This expansion mirrors the development of environmental law and policy, from local regulations addressing immediate pollution concerns to international agreements that acknowledge our shared responsibility for the planet's future.

However, this expansion of moral consideration faces significant psychological barriers. The concept of psychological distancing suggests that events or consequences that are temporally, spatially, or socially distant are perceived as more abstract and less pressing. This explains why the long-term, global nature of air pollution makes it particularly challenging for individuals to prioritize, especially when faced with more immediate concerns. As noted in Chapter 1, London's "Great Smog" of 1952 claimed thousands of lives, yet significant action wasn't taken until the Clean Air Act of 1956. This delay exemplifies our recurring pattern of reactive rather than proactive environmental policy, a pattern that continues to haunt us today. Even when action was taken, the corporate impulse was to shift the problem elsewhere, instead of solving it at the root.

Moreover, the global economic system, built on the principles of comparative advantage and specialization, has created a world where the production of goods (and consequently, pollution) is concentrated in developing countries, while consumption is highest in developed nations. This spatial separation of cause and effect further exacerbates the psychological distance, making it easier for those in developed countries to overlook the environmental costs of their lifestyle. The electronic waste crisis exemplifies this dynamic — while consumers in developed nations enjoy the latest smartphones and laptops, the toxic fumes from informal e-waste recycling sites in places like Agbogbloshie (one of the world's largest e-waste processing sites that was ultimately demolished by Ghana's government in 2021) pollute local air and water resources. To move forward and shape blue skies for all, we must address these inherent aspects of human nature and the global systems they have created. This requires a multi-faceted approach that acknowledges our limitations while striving to transcend them.

Firstly, we must work to bridge the psychological distance that separates individuals from the consequences of their actions. This can be achieved through education and awareness campaigns that make the abstract concept of global air pollution more concrete and immediate. For example, utilizing virtual reality technology to allow people to experience the air quality in polluted cities, or creating more vivid and personal narratives around the health impacts of air pollution, could help overcome the abstractness that often characterizes environmental issues. Secondly, we need to reframe the narrative around economic development and environmental protection. The false dichotomy between economic growth and environmental quality needs to be challenged. By highlighting the economic costs of air pollution, including healthcare expenses and lost productivity, we can align environmental protection with self-interest, making it a more compelling priority for both individuals and nations. Thirdly, we need to tap into other aspects of human nature that can counterbalance our tendency toward self-interest. The human capacity for empathy, for instance, can be a powerful motivator for environmental action when properly engaged. By fostering global connections and emphasizing our shared humanity and shared environment, we can potentially expand our circle of concern to encompass the global community. The success of international climate youth movements has demonstrated how shared values and concerns can transcend national boundaries and create powerful momentum for change.

Many have advocated technological advancement as the ultimate solution to air pollution. Artificial intelligence can help with predictive modeling for air quality and optimization of industrial processes. Blockchain technology enables better environmental impact tracking and carbon credit trading. The Internet of Things allows for real-time pollution monitoring and smart resource management. However, our relationship with technology in addressing such challenges is paradoxical. While modern technology offers numerous solutions to air pollution — from sophisticated air quality monitoring networks to clean energy technologies and carbon capture systems — the production of these very technologies often creates pollution in developing nations. The energy consumption of digital solutions contributes to carbon emissions, and the disposal of obsolete environmental technologies creates new environmental challenges through e-waste. This contradiction is particularly evident in the solar panel industry, where the manufacturing process in countries like China, as well as the recycling of solar waste, often involves significant environmental costs (Atasu *et al.*, 2021), even as the final products help reduce emissions elsewhere.

In Charles Dickens' masterpiece *Bleak House*, London is perpetually shrouded in fog, a metaphorical and literal manifestation of society's moral and environmental decay. The novel opens with an unforgettable description of a city choking on its own industrial progress: "Fog everywhere. Fog up the river... fog down the river... Fog on the Essex marshes, fog on the Kentish heights" (Dickens, 1868, p. 1). The corrupt legal system depicted in the book, symbolized by the case of Jarndyce and Jarndyce, parallels our modern regulatory challenges in addressing air pollution. Just as the Court of Chancery fails its citizens through bureaucratic inertia, many of today's environmental protection agencies struggle with slow-moving legislation and enforcement. The technological response to air pollution faces its own version of what Dickens portrayed as institutional hypocrisy. Major corporations like Apple and Samsung champion environmental initiatives while their supply chains continue to contribute to pollution in developing nations. The character of Jo, the poor street sweeper who dies from an infectious disease fostered by London's unsanitary conditions, finds modern counterparts in the estimated millions of people who die prematurely each year from air pollution-related causes.

In *Bleak House*, the fog finally lifts when truth and justice prevail, though at great cost to many characters. Similarly, our path to cleaner air requires confronting uncomfortable truths about our economic systems and personal choices. Companies like Tesla and Siemens are leading technological transformations in their respective industries, but like the reforms that come too late for Jo in the novel, we must question whether these changes are advancing quickly enough to prevent unnecessary suffering. The novel's central metaphor of fog as both physical pollution and moral blindness speaks powerfully to our current situation. We have the technology to monitor, predict, and reduce air pollution — capabilities far beyond anything Dickens could have imagined — but we still struggle with the same human tendencies toward self-interest and short-term thinking that he so masterfully depicted. As we deploy smart sensors, AI-driven prediction models, and clean energy systems, we would do well to remember *Bleak House's* warning about the cost of institutional and moral failure to address systemic problems before they claim more victims.

The path to blue skies for all requires us to confront the complexities of human nature and the global systems we have created. It demands that we expand our ethical consideration, bridge psychological distances, rouse our moral courage, and realign our economic structures with our environmental goals. This journey will not be easy, as it requires us to transcend some of our most deeply ingrained tendencies. In facing these environmental challenges, we often find ourselves caught between hope and despair, between action and paralysis. The magnitude of the problem can make individual efforts seem insignificant, like fireflies against the darkness of industrial pollution. Yet, transformative change has always begun with individual acts of courage multiplied across communities. Those who can act must act, and those who can speak must speak — whether they are scientists presenting research, community organizers rallying neighborhoods, or workers striking for their future. In a world where environmental injustice often seems insurmountable, like a mountain of prejudice too steep to climb, we might feel ourselves shrinking from the challenge. The politics of pollution control can indeed be soul-destroying, particularly when confronting powerful interests that view environmental concerns as mere obstacles to profit. However, history teaches us that collective resistance, even in its simplest forms, can gradually reshape these mountains of indifference.

Recently, I was attending a session of the *Humanistic Leadership Academy* where the facilitators shared with us a poem by Marge Piercy, "To be of use." I was particularly struck by the following lines (Piercy, 1973):

I love people who harness themselves, an ox to a heavy cart,
who pull like water buffalo, with massive patience,
who strain in the mud and the muck to move things forward,
who do what has to be done, again and again.

The poem's imagery of people who "harness themselves, an ox to a heavy cart" spoke deeply to the kind of dedication our environmental challenges demand. Like the water buffalo pulling with massive patience, we too must persist in our efforts, doing what needs to be done again and again, knowing that lasting change comes not from singular actions but from sustained commitment. It is precisely in this struggle against our own limitations that we find the essence of human progress and the hope for a sustainable future. We must acknowledge the selfish aspects of our nature while striving to cultivate our capacity for collective action and global empathy. In this delicate balance lies the potential for true change — a world where the blue skies above India are as clear as those above any developed nation, and where the air we breathe is a testament not to our divisions, but to our common humanity and shared stewardship of this planet. Just as Piercy celebrates those who strain against the weight of necessary work, I hope that as each of us acts with a little love to protect our environment, harnessing ourselves to this vital task with the patience of water buffalo and the persistence of oxen, the skies will become a beautiful view for people everywhere.

References

Atasu, A., Duran, S., and Van Wassenhove, L. (2021). The Dark Side of Solar Power. *Harvard Business Review*. Retrieved on March 12, 2024 from https://hbr.org/2021/06/the-dark-side-of-solar-power.

Dickens, C. (1868). *Bleak house*. London, UK: Chapman and Hall.

Einstein, A. (n.d.). *Wisdom Quotes*. Retrieved on March 12, 2024 from www.wisdomquotes.com/topics/action/index8.html.

Piercy, M. (1973). *To be of Use*. New York, NY: Doubleday.

Singer, P. (1981). *The Expanding Circle: Ethics and Sociobiology*. New York, NY: Farrar Straus & Giroux.

Glossary

Air Quality Index
A numerical scale used to communicate how polluted the air is and what associated health effects might be. Higher numbers typically indicate worse air quality and greater health risks. Different countries may use varying calculation methods and scales.

Association to Advance Collegiate Schools of Business (AACSB)
An international accreditation organization for business schools, which evaluates institutions on variables like quality of education, curriculum development, strategic management, and innovation.

B Economy
An economic system that prioritizes business models focused on both profit and positive social impact, benefit corporations, and other stakeholder-centered enterprises.

Badgir
A traditional Persian wind tower or wind catcher that provides natural air-conditioning in hot, arid climates.

Basel Convention The Basel Convention on the Control of Transboundary Movements of Hazardous Wastes and their Disposal is an international agreement originally adopted in 1989, and which came into force in 1992. It seeks to regulate the transboundary transfer of hazardous wastes and obliges its signatories to ensure that such wastes are managed and disposed of in a safe manner by following an established set of procedures.

Benefit Corporations A type of for-profit corporate entity that includes a positive impact on society, workers, the community, and the environment in addition to profit as its legally defined goals.

Carbon-neutral A state where the net amount of carbon dioxide and other greenhouse gases released into the atmosphere by an entity is zero, achieved through a combination of emission reduction and carbon offset measures.

Circle of ethical consideration A concept that describes the expanding scope of moral concern, from immediate family to all humans, to animals, plants, and entire ecosystems.

Circular economy An economic system aimed at eliminating waste through the reuse, repair, and recycling of materials, with the goal of creating a sustainable, closed-loop system.

Circular electronics An approach to designing, making, and using electronic devices emphasizing sustainability, reuse, and the elimination of waste throughout the product's entire lifecycle.

Comfort shopping The act of making purchases primarily to relieve stress or negative emotions, often involving items that provide immediate comfort or satisfaction.

Construal level theory A psychological theory that describes how the distance of an object or event affects how people think about it. Abstract, high-level construals are used for distant things, while concrete, low-level construals are used for near things.

Cradle-to-cradle A product design philosophy focused on circularity, such that the product's materials and components can be repurposed or recycled indefinitely.

Defeat device Software or mechanical device designed to circumvent or disable emission control systems during normal vehicle operation while passing emission tests under laboratory conditions.

Electronic waste (e-waste) Encompasses various types of discarded electrical and electronic devices and their parts. The category is very broad, including consumer electronics, household appliances, IT equipment, smart toys, light fixtures, dispensers like vending machines, and surveillance devices that are no longer wanted.

Environmental personhood The concept of granting certain natural entities, such as rivers, forests, or mountains, the rights of a legal person in order to protect them from harm and ensure their preservation for future generations.

Fire-stick farming Indigenous land management practice in Australia using controlled burning to promote ecological health and food production.

Green gentrification The process where environmental improvements to an urban area (such as parks, green spaces, or sustainable infrastructure) lead to increased property values and displacement of lower-income residents.

Green growth	An economic development strategy that emphasizes environmentally sustainable economic progress, focusing on both GDP growth and environmental protection.
Greenhouse gas emissions	The release of gases that trap heat in Earth's atmosphere, primarily including carbon dioxide, methane, nitrous oxide, and fluorinated gases. These emissions come from both natural sources and human activities such as burning fossil fuels, industrial processes, and agricultural practices.
Hybrid knowledge	The integration of traditional ecological knowledge with modern scientific understanding, creating more comprehensive approaches to environmental management and problem-solving.
Hyperbolic discounting	A cognitive bias where people show a preference for smaller, immediate payoffs rather than larger, future ones in a way that is mathematically hyperbolic.
Institutional logics	The socially constructed patterns of practices, assumptions, values, and rules by which individuals and organizations provide meaning to their activities.
Intelligent mine	A mining operation that integrates digital technologies, automation, and data analytics to improve safety, efficiency, and environmental performance.
Kaitiakitanga	A Māori concept representing guardianship and stewardship of the environment, encompassing the responsibility to protect and nurture the natural world for future generations.
Mottainai	A Japanese term expressing regret over waste and a sense of respect for the intrinsic value of resources.

Offshoring	The practice of relocating business processes from one country to another to reduce costs. This can include both outsourcing to third parties and establishing company-owned facilities abroad.
PM2.5	Particulate matter with a diameter of 2.5 micrometers or less. These fine particles are a major air pollutant that can penetrate deep into the lungs and bloodstream, causing significant health problems. Sources include vehicle emissions, industrial processes, and natural phenomena like wildfires.
Privilege of blue skies	The inequitable distribution of air quality, where certain communities enjoy cleaner air while others suffer from pollution. This phrase highlights environmental justice issues and the disproportionate impact of air pollution on disadvantaged communities.
Psychological distancing	The perception of separation between the self and other things, events, or experiences across different dimensions.
Quantum social consciousness	An emerging theoretical framework that applies quantum principles to understanding collective social awareness and behavior, particularly in relation to environmental and social issues.
Renewable Energy Certificates (RECs)	Tradable, market-based instruments representing one megawatt-hour of electricity generated and supplied to the electricity grid from a renewable energy resource. RECs are sold separately from the actual electricity in unbundled certificates.
Reshoring	The practice of bringing manufacturing and services back to the country of origin after they were previously offshored.

Retail therapy	The temporary mood improvement or emotional satisfaction gained from purchasing goods. Retail therapy often involves the purchase of discretionary items and is characterized by the shopping experience itself being therapeutic, regardless of the items purchased.
Satellite city	A smaller metropolitan area located near and economically tied to a larger metropolitan center, but maintaining its own independent identity and government.
Smart city	An urban area that uses various types of electronic data collection and technology to efficiently manage assets, resources, and services.
Smog	A type of air pollution combining smoke and fog, typically resulting from the interaction between sunlight and emissions from vehicles, industries, and other sources.
Social impact	The effect of an activity or initiative on the well-being of a community, including changes in economic, environmental, and social conditions.
Socio-economic status (SES)	Describes the social standing of individuals or groups, typically based on multiple variables like education, income, and occupation.
Techno-nationalist urbanism	An approach to urban development that emphasizes technological sovereignty and national identity through city planning and infrastructure.
Urban renewal project	Large-scale redevelopment programs aimed at modernizing and improving deteriorated urban areas. While often intended to revitalize communities, these projects may raise concerns about displacement, cultural preservation, and environmental justice.

Value of Statistical Life (VSL)	An economic measure used in policy analysis representing the financial value society places on reducing the average number of deaths by one. This metric is often used by decision-makers to evaluate the cost-effectiveness of environmental and safety regulations.
Volatile Organic Compounds (VOCs)	Organic chemicals that easily become vapors or gases at room temperature, emitted by many products and industrial processes. Depending on the constituents, these compounds may contribute to air pollution and can cause various health effects, making them a key concern in indoor and outdoor air quality management.

Index

A

activism, 116–118, 120

air quality, 5–8, 25, 28, 35, 37, 39–40, 48–49, 56, 61–62, 65, 68–69, 73, 76, 84, 95, 106–107, 109–110, 116–120, 126, 128–129, 135–136, 142, 144

air quality index, 33, 39, 65, 118

air quality management, 74–75, 92, 105, 110, 127, 135

air quality monitoring, 60, 75–76, 118, 127

air quality rights, 143, 145–146

artificial intelligence, 45, 75–76

automotive industry, 6, 123–124

B

bamboo, 107–108

Bangladesh, 6–8, 25–26, 30

B Corp movement, 130

B Corporation, 133

B Corps, 135–138, 145

B Economy, 134–135, 137–139

Beijing, 28, 35, 56–58, 60–61, 118, 120

Bitcoin mining, 46

Boston, 88

C

carbon dioxide emissions, 46

carbon emissions, 18

carbon sink, 107

ChatGPT, 45, 47

Chipko movement, 117

circular data centers, 48

circular electronics industry, 48

Clean Air Act, 4, 24, 150

clean air rights, 142

climate, 142

climate change, 30, 36, 43, 62, 68, 71, 111, 133, 142, 144

coal mining, 57

community land trusts, 86–87

corporate–civic partnerships, 127, 128–130

corporate governance, 137–138, 145

corporate responsibility, 138

corporate social responsibility, 127, 135

cryptocurrency mining, 47
crypto mining, 60

D
data center, 43–48
deindustrialization, 8–9
Delhi, 33–34, 44–45, 65–66, 73, 149
digital divide, 78

E
Ekibastuz, 47
Ella Adoo-Kissi-Debrah, 141, 146–147
Ella's law, 141–142
environmental gentrification, 84
environmental injustice, 35, 37, 84,
    87, 91, 93, 95–97
environmental personhood, 128–129
Environmental Protection Agency
    (EPA), 4–5, 24–25, 118
environmental, social, and
    governance (ESG), 134, 137
environmental sustainability, 19, 21,
    76
ESG metrics, 135–136
e-waste, 48

F
fashion industry, 17
fire management, 103–104
fire-stick farming, 103
Foxconn, 6, 26–27

G
generative AI, 76
green city, 92
green corridors, 127
green gentrification, 84–85, 87
greenhouse gas emission, 74, 134,
    144
Greenpoint, 83–84

H
human rights, 142, 144

I
indigenous knowledge, 103, 105,
    109
industrialization, 7, 19, 67, 69
Interface, 124, 135

K
kaitiakitanga, 105

L
London, 3, 141, 150, 152
Los Angeles, 115, 117
Love Canal Homeowners
    Association, 126

M
mass consumption, 8, 55
Mexico City, 117–118, 120
mottainai, 13, 21

N
neem, 104–105, 107
Neom, 92–94, 96–97
New Delhi, 35
Nusantara, 91, 93–94, 96

O
Oakland, 86–87
offshoring, 9, 18, 77

P
particulate matter, 39, 56, 60, 69
Patagonia, 133, 139, 145
Pilsen, 84
Pittsburgh, 126–127
planned obsolescence, 17
pollution-related illnesses, 66

privilege of blue skies, 9–10, 149
psychological distance, 18–19, 47,
    149–151, 153

Q
quantum computing, 45
quantum social consciousness, 127

R
renewable energy, 57–59, 74–75, 93
Renewable Energy Certificates
    (RECs), 44–45
reshoring, 77
robotics, 57, 76–78

S
satellite cities, 92, 95
Seattle, 85–88
Sen, Amartya, 20, 60, 70
Seoul, 85, 92, 96
Shinrin-yoku, 109
shipbreaking, 25–26
shipbreaking industry, 25
smart city, 85, 88, 95

stakeholder capitalism, 137
Stamp Out Smog (SOS), 116–117
steel industry, 4

T
textile industry, 6, 127
traditional ecological knowledge,
    106, 110

V
value of statistical life (VSL),
    24–30
Vancouver, 86
VOC emissions, 143
volatile organic compounds (VOCs),
    143

W
war on pollution, 61, 69
waste, 46
wind towers (badgirs), 106
World Bank, 5, 24–25
World Health Organization (WHO),
    35, 56, 67, 74

www.ingramcontent.com/pod-product-compliance
Lightning Source LLC
Chambersburg PA
CBHW050641190326
41458CB00008B/2368